Armand Marseille Dolls,
Revised
2nd Edition
by
Patricia R Smith

COLLECTOR BOOKS

P.O. Box 3009
Paducah, Kentucky 42001

Cover: 17″ Bisque Oriental baby
with brown sleep eyes
and closed mouth. Marks: A.M./Germany/353/70K. (Author)

The current values in this book should be used only as a guide. They are not intended to set prices, which vary from one section of the country to another. Auction prices as well as dealer prices vary greatly and are affected by condition as well as demand. Neither the Author nor the Publisher assumes responsibility for any losses that might be incurred as a result of consulting this guide.

Additional copies of this book may be ordered from:

Collector Books

P.O. Box 3009

Paducah, Kentucky 42001

@ $9.95 each plus $1.00 postage and handling.

Copyright: Patricia R. Smith, 1981
ISBN: 0-89145-173-o

Printed by Taylor Publishing Company, Dallas, Texas

DEDICATION

This volume of Armand Marseille Dolls is dedicated to the hundreds of collectors who really and truly enjoy adding A.M. marked dolls to their collections.

CREDITS

All photographs by Dwight F. Smith unless noted:

Dolls courtesy of Elaine Boyle: photos by Elaine Boyle

Dolls courtesy of Elizabeth Burke: photos by Gunnar Burke

Dolls courtesy of Barbara Earnshaw: Dwight F. Smith

Dolls courtesy of Jo Fasnacht: photos by Leon Folse

Dolls courtesy of Rosemary Frye: photos by Rosemary Frye

Dolls courtesy of O.D. Gregg: photos by O.D. Gregg

Dolls courtesy of Margaret Gunnel: photos by Dwight F. Smith

Dolls courtesy of Diane Hoffman: photos by Diane Hoffman

Dolls courtesy of Diane Nancy Lucas: photos by Sally Freeman

Dolls courtesy of Lois Milins: photos by Lois Milins

Dolls courtesy of Dorothy Mulholland: photos by Dorothy Mulholland

Dolls courtesy of Mary Partridge: photos by Ted Long.

The following list is of certain Armand Marseille dolls that have given names, and the year they were first introduced. Mold numbers used appear behind the name.

1892:   "Majestic"-some "Majestic" heads were also made by E.U. Steiner.

1893:   "Banker's Daughter" for Butler Bros. (#370)

1894:   "Cleonie" a Black doll.

        "Hindu Boy"

1895:   "Banner Kid Dolls" (#1890/372)

        "Little Aristocrat" for Butler Bros. (#1894)

        "Indian Series" marked A.M. 5/0.

        3200 Series through 1912.

1898:   "Beauty" for W.A. Cissna & Co.

        "Bright Eyes" for W.A. Cissna & Co. (#1897)

        "Lissy" for Butler Bros.

        "Mable" for Butler Bros.

1899:   "Miss Myrtle" for Geo. Brofeldt (#370)

1900:   3700 Series through 1909, which included Fashions and mechanicals.

        "Our Anne" (#370)

        "Alma" for Geo. Borgfeldt.

        "Scowling Indian" (first made in 1890's) for Max Handwerck. (M.H. 2/00)

1901: "Florodora Girl" for Geo. Borgfeldt (who patented "Florodora" in 1921) made in both shoulder plate and socket head.

"Albert" (#1894)

1902: "Marguerite" for Hamberger Co. (#320) (Other companies made this head also)

"Rosebud" for Max Illfelder through 1915.

"Little Sweetheart" for Max Illfelder (#380). Only a few of these heads made by A.M.

1903: "C.M. Bergmann" for C.M. Bergmann.

"My Playmate" for Geo. Borgfeldt.
"My Playmate" heads were made by other companies – S&H, K & R, etc.

1904: "Dollar Princess" for Geo. Borgfeldt.

"Columbia" for C.M. Bergmann, distributed by Louis Wolfe & Co. "Columbia" heads were by A.M. and ran into 1915 with first ones on kid bodies and later in composition.

"Special" for Geo. Borgfeldt.

1905: "Gold Coast Girl" (#370)

"Mabel", "Lilly", "Alma", "Ruth" (to 1914)

"My Princess" (#370)

1906: "Darling Baby" for Strobel & Wilkens.

1908: "My Dearie" for Geo. Borgfeldt (#390/246)

"Pretty Peggy" for Geo. Borgfeldt (#390/245)

"Bernadette" (#640). Some of these heads were made by other companies.

"Fluffy Ruffles" for Stamstag & Hilder (#118)

Some 390 "Jutta" heads for Cuno & Otto Dressel,

"Infant Berry" for Geo. Borgfeldt (#500)

"Bumble Puppy" for Max Illfelder. (#347)

Mold #251, a pouty for Geo. Borgfeldt.

1910:   "Child Berry" (#600)

"Roseland" for Max Illfelder.

"May Queen" (300n)

"Possey" (#390/216)

"Minnit Baby" for Geo. Borgfeldt (#971)

"Sunshine" for Sears & Roebuck.
"Happy Tot" for Geo. Borgfeldt. (#990)

"Queen Louise" heads, but dolls were owned by Louis Wolfe & Co. with bodies made by other companies.

1911:   "My Companion" for Louis Wolfe & Co, and "Daisy" a premium doll offered by Ladies Home Journal and sold by Louis Wolfe & Co. The A.M.'s sold as "Daisy" were second choice for the magazine, as they sold out of the first "Daisy" dolls with heads by Kestner (mold number 171), and heads by Simon & Halbig, with bodies by Handwerck.

"Wonderful Alice" for Geo. Borgfeldt (390)

1912:   "Beatrice" (#700)

"Baby Betty" for Butler Bros.

"Prize Baby" for Geo. Borgfeldt (#326) Some by S & H.

"Rosie Baby" (#326)

"Dorothy" (560a/232)

"Melitta" (incised)

1913: "Missy" lady for Max Handwerck (M.H./A.M.)

"Fany and Dotty" (#251 and 390/240)

"Gibson Girl" for Louis Wolfe & Co. (#400) Marketed also as "Lady".

Beverly Bayne as "Gibson Girl" (#401)

"Miss Millionare" for Sears, Roebuck (370)

"Lilly" for Geo. Borgfeldt.

1914: "Duchess" for Geo. Borgfeldt.

"Jason" for Geo. Borgfeldt (#321)

"Baby Love" (#352)

"New Born Baby" for Louis Amberg & Sons.
#975 line of babies for Louis Wolfe & Co. and Otto Gans, including "Sadie" for Wolfe (#975)

1915: #390 heads for Foulds & Freure, an importer.

#253, a closed mouth googly for Geo Borgfeldt.

"Baby Phyllis" for the Baby Phyllis Doll Co., N.Y.

"Nobbikid" for Geo. Borgfeldt.

"Little Bright Eyes" for Geo. Borgfeldt (#252)

"Little Jane" (#256) for Maar & Sohn

"Peero" for Geo. Borgfeldt (#253) Some made by other companies.

"Louisa" (390n)

"Bernice" Closed mouth, intaglio eyes.

"Little Ann" for Geo. Borgfeldt (#222)

1916: "Friedel" for E.W. Matthes (264)

"Hoopla Girl" for Hitz, Jacobs & Co. (590)

"Little Mary" (#225)

"Educational Doll" for Louis Amberg & Sons (#390/240) Some also made by other companies.

"Lady Marie" for Otto Gans (#970)

"Cheer Ups" a googly for Geo. Borgfeldt (#258)

1917: Fulper Pottery Works borrowed molds from Armand Marseille (and other companies), and by 1918 were producing Fulper heads.

1919: "First Steps" for Louis Amberg & Sons (259)

1920: "Baby Gloria" (Mama voice

"Heidi" for Geo. Borgfeldt (#395)

1921: "New Born Baby" for Louis Amberg & Sons.

"Mobi" for Hermen Schiemer (#917)

1922: "Teenie Weenie" for Geo. Borgfeldt (#362)

"Mimi" for Geo. Borgfeldt (390/216)

"Baby Bobby" for Geo. Borgfeldt (#328)

"Herbie" for Geo. Borgfeldt (991)

"Betsy Baby" for Geo. Borgfeldt (#329)

Some "Kiddiejoy" heads for Hitz, Jacobs & Kassler (#372)

"Wee One" (#351 with rubber body)

1923: "Polly" (#341)

"My Dream Baby" for Arranbee Doll Co. Mold #341-closed mouth and #351-open mouth, and also mold #620 is a form of the Dream Baby. A.M. also made composition "My Dream Baby" with open mouth and two teeth.

"Dora" for Butler Bros. (390/266)

Re-issued "New Born Babe" copyrighted in 1914, for Louis Amberg & Sons, after Geo. Borgfeldt advertised the "Bye-Lo" baby.

1924: "Arranbee" for Arranbee Doll Co.

1925: "Glad Baby" (#325)

"Baby Sunshine" for Louis Wolfe & Co. (#800)

"Ellar" oriental version of the Dream Baby for the Paul Revere Pottery Works.

1926: "Bonnie Babe" for Georgene Averill (All bisque) Most Bonnie Babes made by Alt, Back & Gothschalk.

"Baby Bobby" (#925)

1928: "Just Me" for Geo. Borgfeldt (#310)
       "Clara" (#390/216)

Into 1930's: George Borgfeldt (and other importers) used up dolls from warehouse stock.

GENERAL INFORMATION

Armand Marseille is a very "French" sounding name, and we are sure that it was not the real name of the man that trudged out of the deep valleys of Riga, Russia to settle in Germany. He and his family settled in Koppelsdorf, Thur in 1865, and by the use of brick kilns in his own yard, began a porcelain factory, making table wares.

Within a few years the Marseille factory had grown into a building and employed over 200 people. As the business grew, so did his family. Son, Armand, now known as Herman Marseille, met and married Solveigh Heubach, the sister of Ernest Heubach of Kopplesdorf, maker of doll heads. This new generation of Marseilles began to raise a family of their own, and in 1891 had started producing dolls in the Marseille factory.

The elder Marseille died in the early 1920's, and the son and grandson continued producing more than 1,000 heads a week into the late 1920's with a steady drop to nothing by 1928.

Armand Marseille dolls are found in a great variety, with a great number of "faces" and novelty-type, or character-type heads. The Marseille family was very competitive in the toy market and continued to make the same "dollie" faced dolls for many generations. There are few real character Armand Marseille dolls, and when found, they are a delight to behold.

It is impossible to follow any pattern in Armand Marseille's production for he used the same molds year after year, and with no apparent system of numbering of the molds. Perhaps because the Marseille family entered the doll field "late" (1890's) is the reason their dolls are so highly marked. It seems they were very proud of their dolls, for it would be a rare thing, indeed, to find an unmarked A.M! Another rare item for Armand Marseille dolls is pierced ears, when almost all other German and French makers have a great many pierced ear dolls. The author has only seen or heard of five pierced ear A.M.'s to date.

The Armand Marseille factory mass-produced dolls with thousands of heads firing in the kilns of Walterhausen, Kopplesdorf on any given day. The quality of some A.M. heads did not always equal those of other German makers such as Kestner, nor Simon and Halbig, because of the gigantic output by the artists. The doll industry was a government subsidized business and entire villages were involved with doll and toy making. A great many farm homes had kilns in the yards and would fire for the factories, and many, many people were "artists" in painting heads and bodies. It is suspected that the children were not allowed to paint heads, but many were put to work painting the finger and toe detail on the bodies.

A greater number of A.M. dolls have survived to be on the current market by the sheer fact of ratio to production. They are the dolls most available, and they must be judged, not because they are Armand Marseille's, but on an individual basis of quality. The A.M. output was, at one time, larger than any other doll company, and it is admitted that a great many of the heads produced during this time were inferior by standards used to judge dolls from other firms. It must be noted that a vast amount of A.M. dolls can match equally, or can be far superior to other maker's heads. There are some very beautiful quality A.M. dolls.

Armand Marseille made heads and entire dolls for many other companies and distributors. Among these were: Louis Wolfe & Co., Foulds & Freure, Hitz, Jacobs & Kassel, Butler Bros., W.A. Cissna & Co., Maar & Sohn, Max Handwerck, Louis Amberg & Sons, Otto Gans, Arranbee Doll Co., George Borgfeldt, C.M. Bergmann, Cuno & Otto Dressel, Strobel & Wilkens, Stamstag & Hilder, Max Illfelder, and Herman Schiemer. The Marseilles even had a School of Dollmaking and taught many, including Ernst Reinhardt who learned these methods before coming to the U.S. in 1909 to establish a toy factory in Philadelphia, and then Liverpool, Ohio.

Herman (Armand, Jr.) visited the U.S. and spent eight months studying business methods, and won the Grand Prize at the St. Louis Exposition in 1904 with the C.M. Bergmann doll, "Miss Columbia".

Armand Marseille dolls were distributed into every country, and by 1921 George Borgfeldt had become their sole sales agent. Borgfeldt introduced and continued to handle A.M. dolls even after the A.M. factory had closed in 1928, by using up huge warehouse stock.

The following is a list of known mold numbers used by Armand Marseille, and it must be noted that when the doll is marked with a "390" it means that it is a socket head to be used on a composition body (or shoulder plate). The mold number "370" means that the doll has a shoulder plate on a kid or cloth body.

70, 90, 95, 100, 110, 121, 140, 147, 200, 210, 212, 217, 222, 225, 231, 240, 246, 248, 249, 250, 251, 252, 253, 254, 255, 256, 257, 258, 259, 264, 265, 266, 267, 288, 300n, 301, 310, 320, 320½, 322, 323, 324, 325, 326, 327, 328, 329, 340, 341, 342, 345, 347, 351, 352, 353, 362, 370, 372, 375, 376, 377, 380, 387, 390, 390n, 395, 398, 400, 401, 402, 448, 449, 500, 517, 518, 520, 550, 560, 560a, 580, 590, 600, 620, 640, 690, 700, 701, 750, 753, 760, 800, 853, 854, 917, 920, 925, 957, 966, 970, 971, 975, 977, 980, 985, 990, 991, 992, 995, 996, 997, 1330, 1374, 1776, 1804, 1890, 1894, 1895, 1897, 1899, 1900, 1901, 1908, 1910, 1914, 1921, 2000, 2015, 2549, 2966, 3200, 3300, 3500, 3600, 3700, 3740, 3748, 4008, 83115.

It would be a challenge to locate all the Armand Marseille dolls that have incised names, as the following listing shows: Alma, Baby Betty, Baby Gloria, Baby Phyllis, Beauty, Columbia, Darling Baby, Duchess, Fany, Florodora, Incidein, Heredera, Just Me, Kiddiejoy, Lilly, Lissy, Mabel, Melitta, Majestic, My Companion, My Dearie, My Playmate, Nobbikid, Queen Louise, Rosebud, Roseland, Sadie, Special, Dollar Princess.

Although the "Florodora" dolls do not have a character face, but one considered to be a "dolly" face, they are interesting due to the fact they represent the very famous Florodora Sextette. The 1899-1900 musical comedy "Florodora" (mis-spelled as "Floadora") became the rage of both London and New York due to the fame of the dainty Florodora Sextette and the song, "Tell Me Pretty Maiden". The original members of the sextette were: Margaret Walker, Daisy Greene, Marjorie Relyea, Vaughn Texsmith, Marie Wilson and Agnes Wayburn.

The show opened at the Lyric Theatre in London Nov. 11, 1899 and in New York at the Century Theatre on April 5, 1902. The show was revived (different cast) in New York in 1920, and London in 1915 and 1931.

The word "Florodora" was not the name of the heroine, but was an island in the Philippines owned by a wealthy American, who manufactured a perfume also called Florodora (in plot of play).

Florodora bisque heads came either as a shoulder head or a socket head. All are marked with their name, with the majority being marked on the head, although some will have a label/stamp on the body and the head only marked with the A.M. (for Armand Marseille).

Dolls marked 1892, 1894, 1895, 1900, etc. have been found along with the A.M. mark. At the same time, dolls marked with these dates and a horseshoe, but no A.M. mark must belong to Ernest Heubach. Unless one is found with both the horseshoe and the A.M. mark, we must conclude that they belong to Heubach alone.

A & M initials were used at a porcelain factory in Aich near Karlsbad, Bohemia, by M.J. Moehling (1870-1936) and should not be confused with Armand Marseille, who used the A.M. only, not A & M. There was also a French Marseille, Francis Emile Marseille, from Maisons, Alfort, France (1888). His doll mark was an anchor — There was no relation between these Marseilles.

Many A.M. heads, but not all, show the initials D.E.P., which means "Registered", or D.R.G.M., which means "Incorporated German Company" This means they were patented or registered in Germany.

At one time the word "baby" referred to all dolls. The bent limb baby body was introduced, in composition, in 1909. Also, 1909 brought the flesh colored washable kid bodies, and 1910 started the riveted composition bent baby body. Other riveted types used in the 1890's include the Universal or Ne Plus Ultra joints. 1918 brought the cloth "mama" types of bodies and the toddlers came after 1919.

The German center for making bodies was at Steinach, Thur and the kid was popular in the 1800's and early 1900's, but was used as late as 1930. Jointed kid bodies were made as early as 1842, with riveted joints patented in the 1890's, and here again, we must repeat, used up to 1930.

Composition bodies, first used in the 1860's, became very popular by 1880. There are a few all composition (body and head) dolls made by Armand Marseille.

The Arranbee Doll Co. had Armand Marseille make the heads for their "My Dream Baby" with both open and with full closed mouths. One of the other of these dolls has been referred to as the "Rockabye" baby, and no foundation has been found for this use.

20″ Alma. Shoulder plate. Marks: Alma/3/0. Courtesy Sue Allin.
15″-$150.00 — 20″-$195.00

15" Alma. Shoulder plate. Marks: Alma/9/0. Courtesy Sue Allin.
12"-$110.00 — 20"-$195.00

14″ Alma. Shoulder plate. Marks: Alma/14/0/Germany. Courtesy Sue Allin.

12″-$110.00 — 20″-$195.00

22″ Floradora. Shoulder head. Marks: Floradora/2/0x A.M. Courtesy Sue Allin.

13″-$150.00 — 17″-$175.00

14″ Mabel. Shoulder plate. Marks: Germany/Mabel/12/0. Courtesy Sue Allin.

12″-$145.00 — 26″-$345.00

24″ Queen Louise. Socket head. For Louis Wolfe & Co. in 1910. Marks: 28/Queen Louise/Germany/7. Courtesy Sue Allin.

24″-$300.00

24″ Queen Louise. Socket head. Marks: Queen Louise/Germany/8.
Courtesy Sue Allin.

22″-$275.00 — 29″-$425.00

21″ Socket head. For George Borgfeldt in 1904. Marks: Special/ Germany. Courtesy Sue Allin.

12″-$125.00 — 26″-$345.00

24″ Dollar Princess. Socket head. For George Borgfeldt in 1904.
Marks: The Dollar Princess/b2/Special/Made in Germany. Courtesy
Sue Allin.

12″-$125.00 — 26″-$345.00

7" Socket head. Straight wrists. Marks: Made in Germany/A. 12/0 M. Courtesy Sue Allin.

7"-$95.00

11″ Socket head. Marks: Germany/323/A. 3/0 M. Courtesy Sue Allin.

12″-$650.00

16″ My Dream Baby. Socket head. Made for the Arranbee Doll Co. in 1924. Marks: 341/3K/A.M. Courtesy Sue Allin.

Black 16″-$450.00
White 16″-$325.00

21″ Shoulder plate. Marks 370/A.M. 0½ DEP./Armand Marseille.
Courtesy Sue Allin.

21″-$210.00

19½" Miss Myrtle. Shoulder plate. Mink eyebrows. Marks: D.R.G.M. 374830/3/370 D.R.G.M. 374830/334631 A.M. 2/0x DEP./Made in Germany. Courtesy Sue Allin.

19½"-$225.00

19″ Shoulder plate. Marks: 370/A.M. 2/0x DEP./Made in Germany
Courtesy Sue Allin.

19″-$210.00

24″ Shoulder plate. Marks: 370/A.M. 2½ DEP./Made in Germany. Courtesy Sue Allin.

24″-$225.00

16″ Shoulder plate. Marks: 370/A.M./3/0/DEP./Armand Marseille. Courtesy Sue Allin.

16″-$165.00

17″ Shoulder plate. Marks: 3F0/A. 4/0 M./ Made in Germany. Courtesy Sue Allin.

17″-$170.00

16″ Shoulder plate. Marks: 370/A.M. 5/0 DEP./Made in Germany.
Courtesy Sue Allin.

16″-$165.00

18″ Shoulder plate. Marks: A.M./370/A.M. 5/0 DEP./Made in Germany. Courtesy Sue Allin.

18″-$175.00

17″ Shoulder plate. Marks: 370/A.M. 5/0 DEP./Armand Marseille/ Made in Germany. Courtesy Sue Allin.

17″-$170.00

15″ Shoulder plate. Marks: 3F0/A.M. 6/0 DEP. Courtesy Sue Allin.

15″-$150.00

15½" Socket head. Straight wrists. Marks: Made in Germany 390/A. 0½ M. Courtesy Sue Allin. 15½"-65.00.

16″ Socket head. Straight wrists, original clothes. Marks: Made in Germany/Armand Marseille/390/A. 1 M. Courtesy Sue Allin.

16″-$165.00

14″ Socket head. Marks: Armand Marseille/Germany/390n/A. 4/0x M. Courtesy Sue Allin.

14″-$200.00

26″ Wonderful Alice. Socket head. Fur eyebrows. Marks: D.R.G.M. 377439/Made in Germany/D.R.G.M. 374830/374831/390/A. 11 M. Courtesy Sue Allin.

26″-$285.00

23″ Socket head. Marks: Made in Germany/Armand Marseille/390/ D.R.G.M. 245/1/A. 5 M. Courtesy Sue Allin.

23″-$275.00

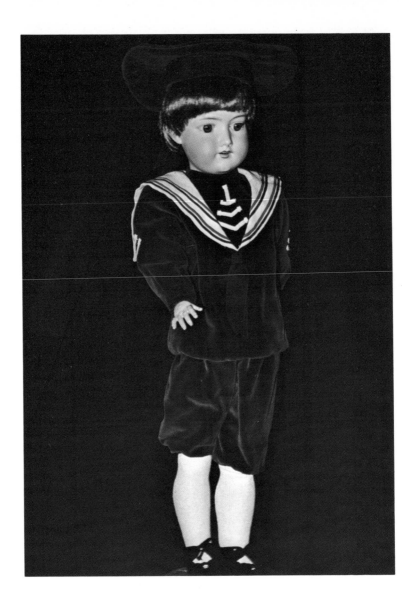

24″ Socket head. For George Borgfeldt in 1904. Marks: Made in Germany/Armand Marseille/390/D.R.G.M. 266/1/A. 6½ M. Courtesy Sue Allin.

24″-$275.00

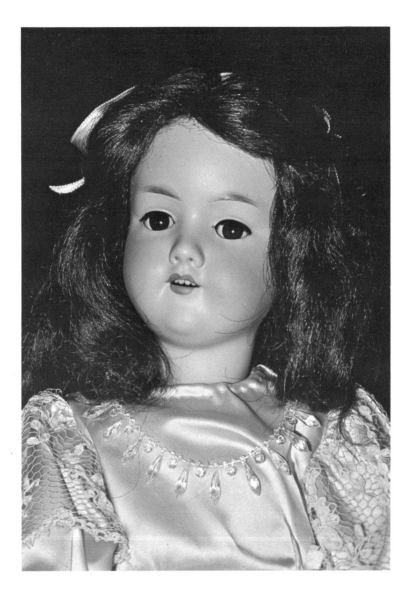

24″ My Dearie. Socket head. 1908-1922. Marks: 390/A.M. 246/1/A. 7 M. Courtesy Sue Allin.

24″-$275.00

9″ Socket head. Marks: Made in Germany/Armand Marseille/560a/
A. 8/0 M./D.R.M.R. 232. Courtesy Sue Allin.

9″-$185.00 — 15″-$350.00

11″ Socket head. Marks: Germany/A. 985 M./410. Courtesy Sue Allin.

11″-$200.00 — 26″-$550.00

17″ Socket head. Marks: Armand Marseille/Germany/995/A. 4½ M.
Courtesy Sue Allin.

12″-$200.00 — 17″-$325.00

19″ Shoulder plate. Turned. Marks: 3200/A.M. 1 DEP. Courtesy Sue Allin.

19″-$265.00

20″ Bisque socket head on fully jointed composition body. (Also came as a shoulder head on a kid body). Open mouth, cheek dimples and sleep eyes. Above average quality bisque. Marks: Baby Betty, in circle DRGM/A.M. Courtesy Elizabeth Burke.

20″-$275.00 — 26″-$325.00

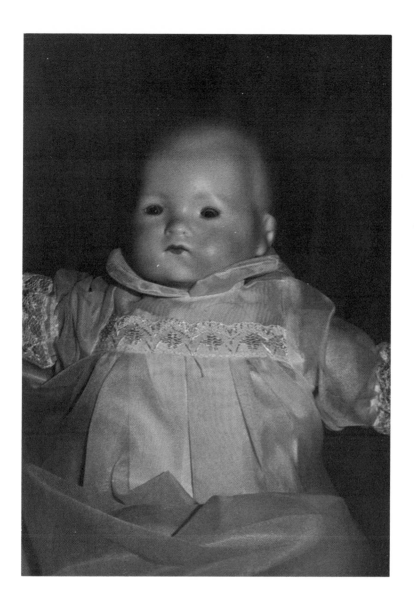

Marks: A.M./Germany. Courtesy Jay Minter.

16″-$325.00

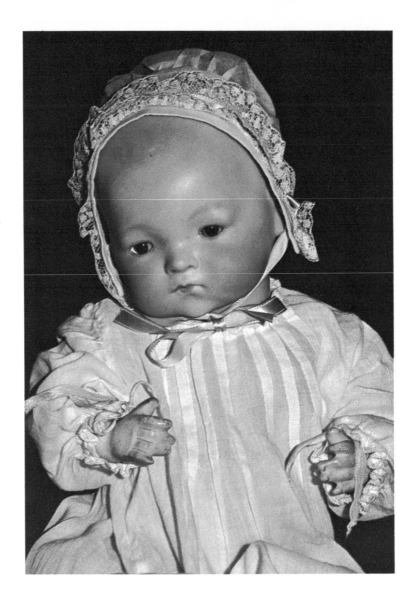

16″ Socket head. Marks: Germany/3/H/41K/A.M. Courtesy Jay Minter.

16″-$325.00

12″ Socket head. Marks: Germany/323/A. 6/0 M. Courtesy Jay Minter.

12″-$650.00

12″ Socket head. Marks: Germany/323/A. 4/0 M. Courtesy Jay Minter.

12″-$650.00

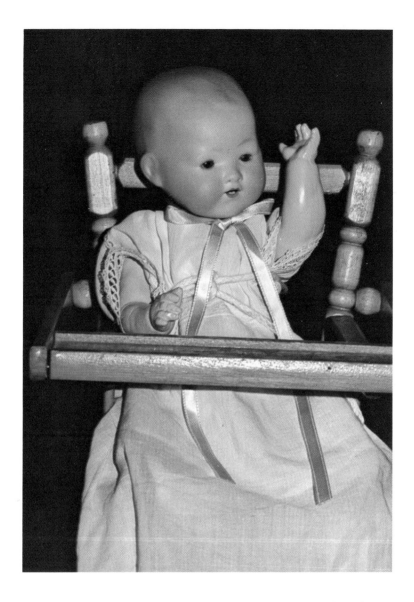

My Dream Baby, Socket head. Marks: A.M. Germany/351/2½K.
Courtesy Jay Minter.

7″-$150.00 — 14″-$300.00

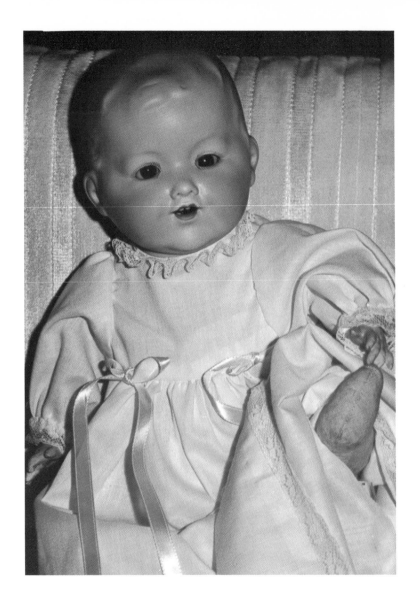

Socket head. Marks: A.M./Germany/352. Courtesy Jay Minter.
12″-$200.00 — 18″-$325.00

25" Socket head. Marks: Germany/390/A. 9 M. Courtesy Jay Minter.

25"-$275.00

24″ Socket head. Marks: Made in Germany/Armand Marseille/560/ A. 5 M./D.R.G.M. Courtesy Jay Minter.

18″-$425.00 — 25″-$600.00

22″ Shoulder plate. Marks: No. 3500/A.M. 7 DEP. (All Cursive)/ Made in Germany. Courtesy Jay Minter.

22″-$300.00

19½" Florodora marked on body in paper sticker with banner and flowers. Head is marked: 370/A.M. — 4-DEP/Armand Marseille. She also has written in black ink her original price of $2.25. Kid body with fat legs. Sleep blue eyes and open mouth. Courtesy Nancy Lucas.

19"-$185.00 — 23"-$275.00

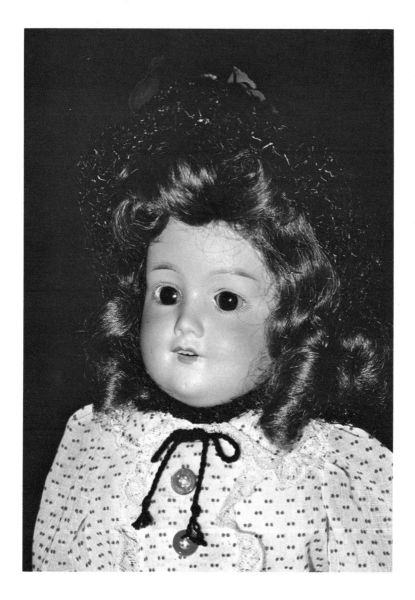

17" Baby Betty. Shoulder plate. For Butler Bros. in 1912. Marks: Baby/0½/Betty/D.R.G.M.

20"-$275.00 — 26"-$325.00

7″ Socket head. Fully jointed. Marks: Made in Germany/12/0.
7″-$95.00

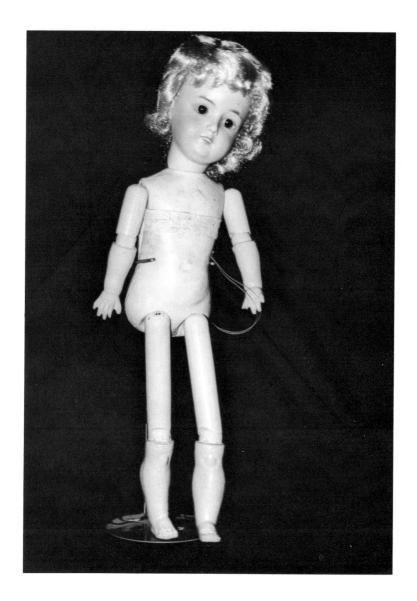

17″ Socket head. Talker. Pull string to say "Mama". Marks: 390/A. 2 M.

17″-$185.00

23″ Socket head. Marks: Armand Marseille/Germany/390/A. 7 M.
23″-$275.00

16½″ Marked: 1894/A.M. 3 DEP. Has wooden ball jointed body with very long upper legs. Blue set eyes, open mouth with four teeth. Courtesy Nancy Lucas.

11″-$135.00
16½″-$210.00
23″-$285.00

16″ Alma. Shoulder plate. Marks: Alma/10/0/Germany. Courtesy Kathy Walters.

17″-$165.00 — 26″-$250.00

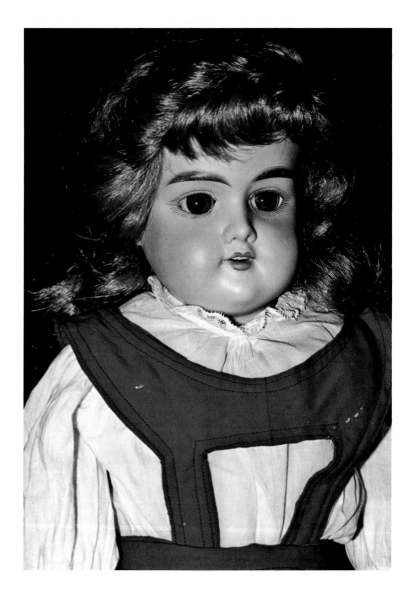

20″ Shoulder head. Made for W.A. Cissna & Co. in 1898. Marks: A.M./Beauty/Made in Germany. Courtesy Kathy Walters.

20″-$250.00 — 26″-$345.00

24″ Duchess. Socket head. Marks: Duchess/A.8M./Made in Germany. Courtesy Kathy Walters.

12″-$90.00 — 24″-$310.00

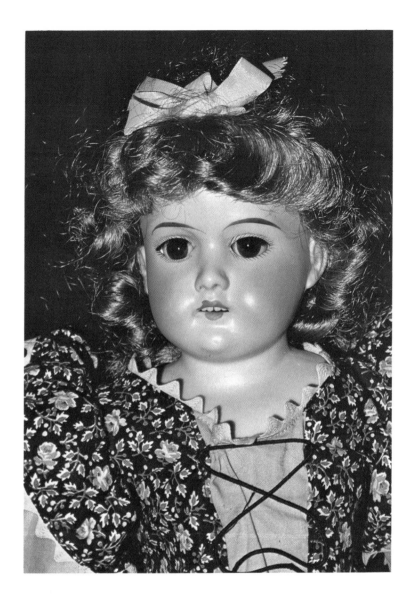

25" Floradora. Shoulder plate. Turned. Marks: Floradora/A.O.M./
Made in Germany. Courtesy Kathy Walters.

18"-$185.00 — 23"-$275.00

20″ Floradora. Socket head. Marks: Made in Germany/Floradora/
F4M. Courtesy Kathy Walters.

17″-$175.00 — 25″-$325.00

16″ Floradora. Shoulder plate. Turned. Marks: Floradora/A. 4/0 M./Made in Germany. Courtesy Kathy Walters.

16″-$170.00 — 25″-$325.00

25″ Floradora. Shoulder plate. Marks: Floradora/A. 6 M./Made in Germany. Courtesy Kathy Walters.

16″-$170.00 — 24½″-$325.00

12″ Floradora. Shoulder plate. Marks: Floradora/Armand Marseille/ A. 7/0 M./ Made in Germany. Courtesy Kathy Walters.

12″-$135.00 — 15″-$165.00

25″ Floradora. Socket head. Marks: Made in Germany/Floradora/A.
77 M. Courtesy Kathy Walters.

14″-$160.00 — 27″-$365.00

18″ Mabel. Shoulder plate. Marks: Germany/Mabel/310. Courtesy Kathy Walters.

14″-$150.00 — 25″-$325.00

24″ Queen Louise. Socket head. Made for L. Wolfe & Co. in 1910.
Marks: Queen Louise/100/Germany. Courtesy Kathy Walters.

24″-$300.00

14″ Shoulder plate. Marks: A.M.-6/O-DEP./Armand Marseille/ Made in Germany. Courtesy Kathy Walters.

12″-$110.00 — 26″-$250.00

13" Shoulder plate. Marks: Germany/A.M. 8/0. Courtesy Kathy Walters.

13"-$125.00

20½" Shoulder plate. Marks: (Scroll) A.M. 95-5. Courtesy Kathy Walters.

20"-$195.00

12″ My Dream Baby, Socket head. Made for Arranbee Doll Co. in 1924. Marks: A.M./Germany/341/OK. Courtesy Kathy Walters.

12″-$285.00

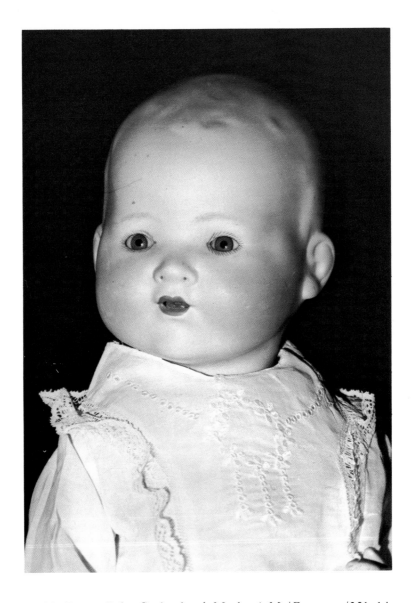

16″ My Dream Baby. Socket head. Marks: A.M./Germany/351. 14.
K. Courtesy Kathy Walters.

16″-$350.00

17½" Shoulder head. Marks: Armand Marseille/370/A.D.M./Made in Germany. Courtesy Kathy Walters.

17½"-$175.00

22″ Shoulder plate. Marks: 370/A.M. 4 DEP./Armand Marseille/
Made in Germany. Courtesy Kathy Walters.

22″-$225.00

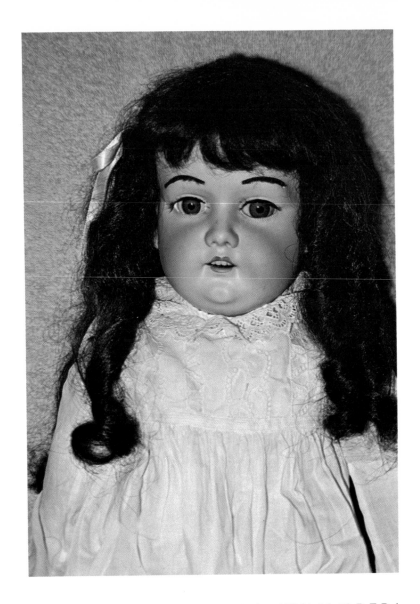

25″ Shoulder head. Fur eyebrows. Marks: 370/A.M.-10-D.E.P./
Armand Marseille/Made in Germany. Courtesy Kathy Walters.

25″-$275.00

12" My Dearie. Socket head, stick legs. Marks: Made in Germany/
Armand Marseille/390/D.R.G.M. 246/1/A. 6/0 M. Courtesy Kathy
Walters.

12"-$120.00

15″ Socket head. Marks: 1894/A.M.1 DEP./Made in Germany.
Courtesy Kathy Walters.

15″-$185.00

17″ Shoulder plate. Marks: 1894/A.M. 9/0 DEP. Courtesy Kathy Walters.

17″-$210.00

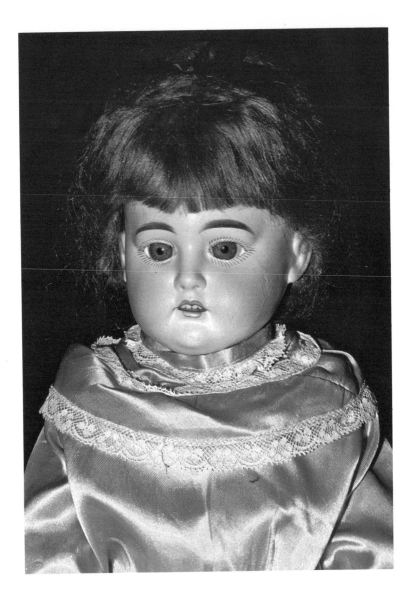

21" Bright Eyes. Shoulder plate. Marks: 1897/A.M. 5 DEP. Courtesy Kathy Walters.

21"-$255.00

14″ Shoulder Plate. Marks: E/3200/A.M. 6/0 DEP./Made in Germany. Courtesy Kathy Walters.

14″-$195.00

20″ Socket head with open mouth, sleep eyes and on a fully jointed composition body. Marks: 1894/A.M. DEP./Made in Germany/5. Courtesy Margaret Gunnell.

13″-$165.00 — 20″-$250.00

14″ Shoulder plate. Turned. Felt body. Marks: 3200/A.M. 3/0 DEP. Courtesy Dorothy Westbrook.

14″-$195.00

13″ cloth with celluloid hands. Original clothes. Marks: Germany/ 370/A 9/0 M. Courtesy Leslie White.

13″-$125.00

14" My Dream Baby. Marks: 351/12/0x. Courtesy Donna Hicks.
8"-$165.00 — 14"-$315.00

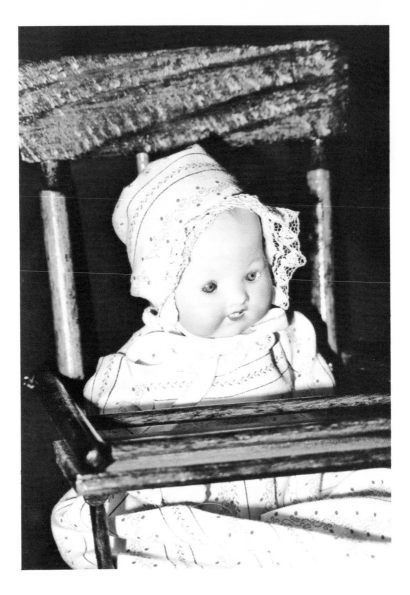

8″ My Dream Baby. Socket head. Marks: A.M. 351/16/0K. Courtesy Donna Hicks.

6″-$135.00 — 10″-$235.00

18″ Patrice. Socket head. Marks: 390n/A. 2½ M. Courtesy Donna Hicks.

18″-$225.00

16" Bisque headed, two-faced baby. Head rotates to show other face. This is head of Armand Marseille "My Dream Baby" made for the Arranbee Doll Co. Closed mouth. Courtesy Rosemarie Frye.

16"-$550.00

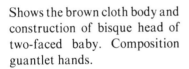

Shows the brown cloth body and construction of bisque head of two-faced baby. Composition guantlet hands.

Shows second face of two-faced baby. Has painted eyes and molded tears. Mouth in open/closed with painted teeth. Courtesy Rosemarie Frye.

21" Bisque shoulder head on kid body with bisque lower arms. Open mouth and has talker mechanism in head. Marks: A3M/DRGM 201013, on shoulder plate. Made by Armand Marseille (Germany). Courtesy Mary Partridge. Photo by Ted Long.

16"-$295.00
21"-$450.00

Shows the talking lever of the DRGM 201013 Armand Marseille doll. Courtesy Mary Partridge. Photo by Ted Long.

19″ Toddler marked: A. 985 M./Germany/8. On toddler composition body that is fully jointed. Sleep eyes and open mouth. Courtesy Margaret Gunnell.

12″-$200.00 — 19″-$400.00

27" Socket head with sleep eyes/lashes and open mouth. Composition, fully jointed body. Marks: 390n/D.R.G.M. 246/1/A. 11 M. Courtesy Margaret Gunnell.

12"-$170.00 — 18"-$225.00 — 27"-$350.00

23" Queen Louise. Socket head on fully jointed composition body. Sleep eyes/lashes and open mouth. Marks: 305/Queen Louise/ Germany. Queen Louise of Sweden is closely related to the royal family of England. She is the second cousin of Queen Elizabeth, her brother is Lord Louis Mountbatten, and she is the aunt to the Duke of Edinburgh. She was married to the then Crown Prince Gustavus Adolphus of Sweden. Courtesy Margaret Gunnell.

15"-$165.00 — 23"-$300.00

24" Queen Louise on fully jointed composition body. Sleep eyes and open mouth. Marks: 29/Queen Louise/100/Germany. Courtesy Margaret Gunnell.

24"-$300.00 — 29"-$425.00

31″ tall, large beautiful bisque Armand Marseille that is marked A. 14 M. Fully jointed composition body, and open mouth. May be original wig. Courtesy Diane Hoffman.

31″-$495.00 — 36″-$650.00 — 40″-$900.00

12″ Armand Marseille googly with mold marks: 200//A. 3/0 M.//
Germany//DRGM//243. All original clothes and wig. Jointed
mache and composition body. Set brown eyes to the side and closed
mouth. (Author)

12″-$995.00

15″ Baby with bisque head, sleep eyes and open/closed mouth with molded tongue. Head marked: 251 G.B./Germany/A 1M DRGM 248. Five piece bent leg baby body of composition. 1st Place Winner, Regional Convention, Wichita 1978. Courtesy Barbara Earnshaw.

12″-$350.00 — 15″-$425.00

10″ Googlie that is marked: G 253 B/A 5/0 M DRGM. Made by Armand Marseille and distributed through the George Borgfeldt Company. Original fur and boots on five piece body. These "snow babies" became very popular after Admiral Byrd discovered the North Pole. Courtesy O.D. Gregg.

10″-$795.00

8″ "Just Me" with painted bisque head and on five piece composition body. Original clothes and wig. Marks: Germany Registered/A 310/11/0 M. Many of these small "Just Me" dolls were dressed by Vogue Doll Co. and bear the Vogue tag. Courtesy O.D. Gregg.

8″ Painted bisque-$250.00 — 8″ Fired in color-$600.00

22" Bisque socket head on five piece, bent leg baby body. Painted upper and lower lashes, sleep eyes with real lashes, open mouth with two upper teeth. Mache body. Marks: Germany/G 327 B/DRGM 259/A 12 M. It has been noted by Dorothy Coleman that the G.B. may well stand for Gabriel Brenda and not George Borgfeldt. Courtesy Diane Hoffman.

22"-$465.00 — 26"-$550.00

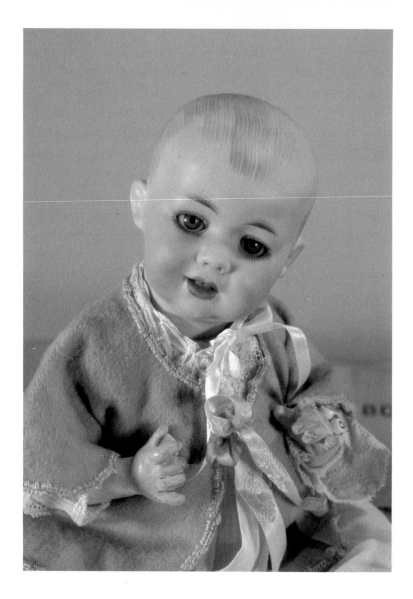

14″ with 8½″ head circumference bisque head baby on five piece bent limb body. Sleep eyes, open mouth, and brush stroke hair. Marks: Germany/G. 326 B/A. 1. M./D.R.G.M. Two lower teeth. Marketed in Germany as "Rosie Baby", and imported by George Borgfeldt as "Prize Baby"/"1st Prize Baby" in 1912. (Author)

12″-$250.00 — 14″-$285.00 — 16″-$325.00

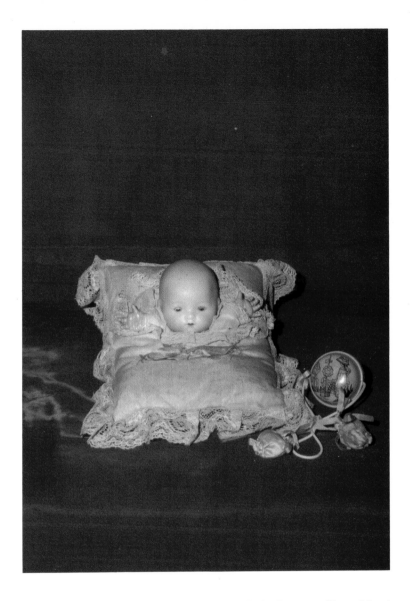

18″ Bisque head, celluloid hands and body built onto pillow. Hand inserted in back makes it a puppet. Blue sleep eyes, open mouth and marked: A.M./Germany/341-10. The celluloid rattle has a pink and blue frozen bisque and says: "Mary, Mary Quite Contrary." Courtesy Jo Fasnacht.

11″-$285.00 — 18″-$425.00

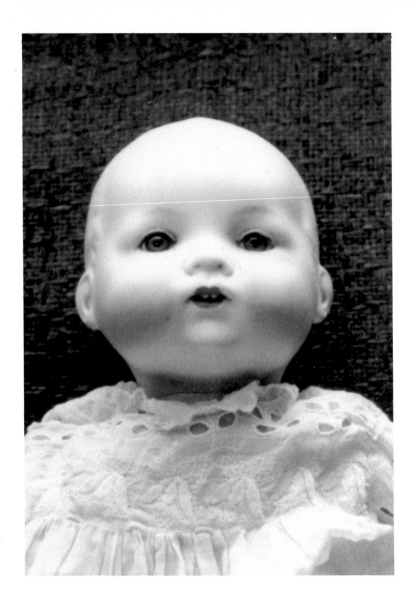

17″ All cloth body, bisque head and composition arms (lower). Has music box in stomach and when pressed plays "Home Sweet Home". Marks: 342/4. Doll has a socket head and shoulder plate. Open mouth with molded teeth. Courtesy Lois Milius.

17″-$425.00

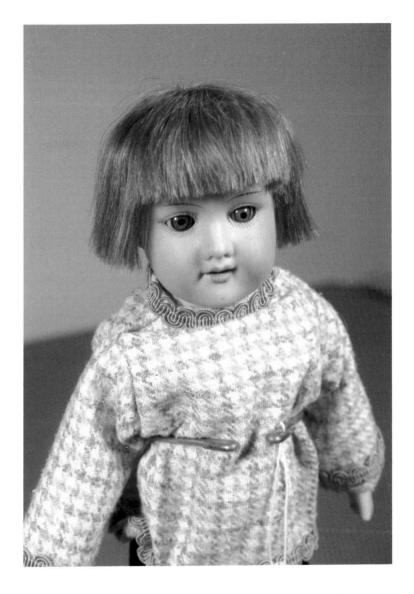

15" Shoulder head on all kid body with bisque lower arms. Sleep eyes and open mouth. Marks: Germany/370/A. 5/0 M. Courtesy Margaret Gunnell.

15"-$150.00 — 29"-$350.00

16″ Marked: 370/A.M. 0½ DEP. Shoulder head with all kid body and bisque lower arms. Green sleep eyes and open mouth. Has molded eyebrows. Courtesy Margaret Gunnell.

16″-$165.00 — 26″-$285.00

17" Shoulder head on all kid body with bisque lower arms. Open mouth and set eyes. Marks: Germany/370/A. 3/0 M. Courtesy Margaret Gunnell.

12"-$120.00 — 17"-$170.00

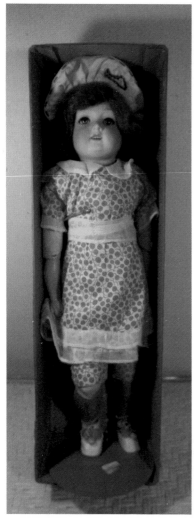

19½" doll in original 21½" box. Bisque shoulder plate on jointed kid body with composition lower arms and lower legs. Short fly-a-way eyebrows, sleep eyes, painted upper and lower lashes with eyelashes. Ears not pierced. All original except hat, which was made by original owner. Canvas type shoes with white pom poms. Mohair wig. Marks: A 3 M/370. Box: Dainty Dorothy. Courtesy Diane Hoffman.

19½"-$225.00 — 19½" in box-$375.00

Shows the end of the box on the Dainty Dorothy doll. Courtesy Diane Hoffman.

22″ Shoulder head with sleep eyes/lashes and open mouth. All kid body with bisque lower arms. Marks: 370/A.M. 2 DEP./Armand Marseille/Germany. Courtesy Margaret Gunnell.

16″-$165.00 — 22″-$225.00

7″ twins marked: Germany/390/A 3/0 M. Sleep blue eyes, open mouth with teeth and mohair wigs. Very well modeled bodies that are in five pieces with painted and molded on shoes with high heels, and painted on silk stockings. Courtesy Dorothy Mulholland.

7″ $85.00

10″ Painted bisque head on five piece composition body. Clothes appear to be original with underwear sewn onto dress. Two gold medal bracelets, one with chain has a heart with a purple stone. Mohair wig, blue sleep eyes and open mouth with four upper teeth. The shoes and hose are painted on. Courtesy Nancy Lucas. Doll is marked: Germany/390/A.9/0 M.

10″-$75.00

13½″ boy on crude, stick type body with torso that looks like pressed wood. Body and limbs are made in five pieces. Marks: Made in Germany/Armand Marseille/390. Original clothes except shoes and hat. Sleep eyes and open mouth. Courtesy Nancy Lucas.

13½″-$140.00 — 16″-$165.00

21″ Bisque head on pasteboard and wood body with stick limbs. Open mouth and sleep eyes. Marks: A.M. 390/Germany. Courtesy Elaine Boyle.

21″-$200.00 — 30″-$400.00

22″ Socket head on fully jointed composition body. Sleep eyes and open mouth. Marks: Armand Marseille/Germany/390/A. 7½ M. Courtesy Margaret Gunnell.

22″-$210.00 — 29″-$400.00

22″ Marks: A.M./Germany/390/A.6 M. Fully jointed composition body. Open mouth and sleep eyes. Courtesy Margaret Gunnell.

10″-$95.00 — 22″-$210.00

15″ "Barry" with bisque head, closed mouth and painted hair. On fully jointed composition body. Marks: A.M. 500. Courtesy Elaine Boyle.

15″-$1,000.00 — 24″-$1,900.00

20″ Girl version of "Barry" with head of wax over papier mache. Sleep blue eyes and closed mouth. On five piece excellent quality composition body. (Author)

20″-$450.00

HOW TO USE PRICE GUIDE

The prices in this book are based on a perfect, well-dressed doll. This means a doll that is ready to go straight into a collection without a "hospital" stop, or a stop at the seamstress.

When you start to buy an Armand Marseille doll, first look at the quality of the bisque. Is it smooth and pale? Are the features painted well? Are the eyes straight? Are the teeth present and straight? Then check to see that the head has no hairline cracks. Check the body to see that it is in good condition and is the correct style.

Never buy, or sell, any antique doll without first completely undressing it and checking everything about it.

Again, the prices in this book are based on a perfect, well-dressed Armand Marseille doll. Original clothes or box will add to these prices.

A damaged doll (hairline cracks, chips, breaks, cracked shoulder plates, eye chips, missing fingers, or broken kid and missing limbs) will bring a lot less than prices shown in this book.

In the following pages of this guide, all dolls have glass eyes and open mouths unless noted.

PRICE GUIDE

MARKS	DESCRIPTION AND PRICE
ALMA 3/0	Shoulder head. 12" $110.00 17" $165.00 26" $250.00
ALMA 9/0	Shoulder head. 12" $110.00 15" $150.00 26" $250.00
ALMA 10/0 GERMANY	Shoulder head. 12" $110.00 26" $250.00
ALMA 14/0	Shoulder head. 12" $110.00 20" $195.00 26" $250.00
Germany	
MADE IN GERMANY BABY 0 BETTY D.R.G.M.	Made for Butler Bros in 1912. Shoulder head and socket head. 12" $135.00 17" $200.00 26" $325.00
MADE IN GERMANY BABY 0½ BETTY D.R.G.M.	Shoulder and socket head. 12" $135.00 20" $275.00 26" $325.00
BABY GLORIA GERMANY 3	Shoulder head. 16" $450.00
BABY PHYLLIS 4 240 MADE IN GERMANY	For Baby Phyllis Doll Co. of N.Y.C. Cloth body. 1915. 12" $225.00 23" $375.00

A.M.
BEAUTY
MADE IN GERMANY

For W.A. Cissna Co. in 1898.
Shoulder head.
12" $95.00 20" $250.00 26" $345.00

BONNIE BABE CORP.
GEORGINE AVERILL
GERMANY
11

For Georgine Averill in 1926.
All bisque.
6½" $475.00 8½" $550.00

COLUMBIA
MADE IN GERMANY

For C.M. Bergmann Co.
Shoulder head. 1904.
24" $325.00

A.M.
DARLING BABY

1906 for Strobel & Wilkens.
12" $145.00 18" $265.00

DUCHESS GERMANY
A. 6 M.

For Geo. Borgfeldt in 1914.
12" $90.00 24" $310.00

DUTCHESS
A 8 M

Socket head.
12" $90.00 26" $365.00

MADE IN GERMANY
D.R.G.M. 248
FANY
A. 2/0 M.

Socket head. Closed mouth
character baby/toddler. 1913.
14" $1,300.00 22" $2,500.00

FLORADORA
MADE IN GERMANY
A.0 M.

Socket head
13" $150.00 17" $175.00 22" $275.00

FLORADORA
A. 1 M.
MADE IN GERMANY

Shoulder head.
20" $245.00

FLORADORA
A. 0½ M.
GERMANY

Socket head.
17" $175.00 25" $325.00

FLORADORA
A. 2/0x M.
MADE IN GERMANY
D.R.P.

Shoulder head, hair
eyebrows.
19½" $245.00 26" $345.00

FLORADORA
2/0x
A.M.

Shoulder head.
21½" $275.00

FLORODORA
A. 2/0x M.
MADE IN GERMANY

Closed dome. Socket head.
19½" $245.00

MADE IN GERMANY
FLORODORA
A. 2 M.

Shoulder head.
22" $275.00

FLORODORA
A. 2½ M
MADE IN GERMANY
D.R.P.

Shoulder head. Closed
dome. 22½" $275.00

MADE IN GERMANY
FLORODORA
A. 2½ M

Socket head.
19½" $245.00

FLORODORA
A 3 M
MADE IN GERMANY

Shoulder head.
23" $275.00

MADE IN GERMANY
ARMAND MARSEILLE
FLORODORA
A. 3/0 M

Shoulder head and socket.
18" $185.00

FLORODORA
A. 4 M.

Socket head.
23" $275.00

MADE IN GERMANY	
FLORODORA A. 4 M	Shoulder head. 23″ $275.00
MADE IN GERMANY FLORODORA A. 4 M.	Shoulder head. 23″ $275.00
MADE IN GERMANY FLORODORA A. 4 M.	Socket head. 21″ $265.00
FLORODORA A. 4/0 M. MADE IN GERMANY	Turned shoulder head. 16½″ $175.00
FLORODORA A. 5/0 M. GERMANY	Shoulder head. 16″ $170.00
FLORODORA A. 5. M GERMANY	Shoulder head. 24″ $315.00
FLORODORA A.M. 5½ D.R.P. MADE IN GERMANY	Shoulder head. Hair eyebrows. 24½″ $325.00
A. 6 M. FLORODORA DRGM 3777301/377731 MADE IN GERMANY	Shoulder head. Some with hair eyebrows. 25″ $325.00
A.M. 6 DRP FLORODORA	Shoulder head. 26″ $345.00
FLORODORA A. 6. M. MADE IN GERMANY	Shoulder head. 25″ $325.00
FLORODORA A. 6 M.	Socket head. 21″ $265.00
MADE IN GERMANY FLORODORA A. 7 M.	Socket head. 22″ $275.00
FLORODORA A. 7/0 M. MADE IN GERMANY	Shoulder head. 14″ $160.00
FLORODORA A.M. 7 D.R.P. GERMANY	Shoulder head. Hair eyebrows. 26″ $345.00
FLORODORA A. 8. M.	Shoulder head. 28″ $395.00
FLORODORA A. 8/0 M MADE IN GERMANY	Shoulder head. 13″ $145.00
MADE IN GERMANY FLORODORA A. 9/0 M.	Socket head. 12″ $135.00
MADE IN GERMANY FLORODORA A. 11 M.	Socket head. 27″ $365.00
16/0 GERMANY BODY: FLORODORA GERMANY	Shoulder head. 6″ $125.00

MADE IN GERMANY
FLORODORA
A. 77 M.

Socket head.
15" $165.00

1374 FLORODORA
A. 6 M.
D.R.G.M. 3748 30

Shoulder head. Fur
eyebrows.
21" $275.00

1374 FLORODORA
A.G.M.
D.R.G.M. 3748 30

Shoulder head. Fur
eyebrows.
26" $345.00 30" $425.00

INCIDEIN
A. 2½ M.
ARMAND MARSEILLE
MADE IN GERMANY

Shoulder head. 1903.
Reg. #58176. 17½" $200.00

HEREDERA
DRGM 374830 37487
MADE IN GERMANY
A. 2½ M

Shoulder head.
17½" $200.00

JUST ME
REGISTERED
GERMANY
A. 310 3/0 M.

For George Borfelt in 1928.
Closed mouth. Googly.
Socket head.
7" $700.00 12" $1,100.00

JUST ME
REGISTERED
GERMANY
A. 310 M.

Painted bisque socket
head. Googly.
7" $275.00 12" $495.00

JUST ME
REGISTERED
GERMANY
A. 310 6 M.

Socket head googly for
George Borgfeldt in 1928.
10" $900.00 12" $1,100.00
10" Painted bisque: $395.00

JUST ME
REGISTERED
GERMANY
A 310 7 M.

Socket head. Googly.
10" $900.00 13" $1,200.00

JUST ME
REGISTERED
GERMANY
A 310 8 M.

Socket head. Molded hair.
12" $1,100.00 18" $1,700.00

JUST ME
REGISTERED
GERMANY
A 310 11/0 M.

Socket head for George
Borgfeldt in 1928.
Googly.
7" $700.00 12" $1,100.00

GERMANY
KIDDIEJOY
A. O M.

Shoulder head for Hitz,
Jacobs & Kassler in 1922.
9" $165.00 18" $400.00

GERMANY KIDDIEJOY
A.M. 1/0
372

For Hitz, Jacobs & Kassler in
1926. Designed by J.I. Orsini.
Flange neck. Cloth body.
Open/closed mouth. Baby.
9" $165.00 18" $400.00

GERMANY
KIDDIEJOY
372
A. 1. M.

Shoulder head. Closed
dome baby. 1926.
9" $165.00 18" $400.00

GERMANY
KIDDIEJOY
3/0

Shoulder head. Cloth body.
Closed mouth.
Molded hair. Girl.
16" $450.00 20" $750.00

GERMANY KIDDIEJOY A. 4 M.	Flange neck. Cloth body. Molded hair. Closed mouth. Girl. 20" $750.00
GERMANY KIDDIEJOY A.M 375/6	For Hitz, Jacobs & Kassler. Open mouth. 9" $165.00 18" $400.00
LILLY 2/0 MADE IN GERMANY	Shoulder head. For Geo. Borgfeldt, 1913. 17" No price on original.
LILLY 4/0 MADE IN GERMANY	Shoulder head. For Geo. Borgfeldt, 1913. 17" $175.00
LISSY O MADE IN GERMANY	For Butler Bros. Shoulder head. 1898. 20" $225.00
GERMANY LISSY 1	For Butler Bros. Shoulder head, 1898. 20" $225.00
MABEL GERMANY	For Butler Bros. Shoulder head. 1898. 12" $145.00 17½" $200.00
GERMANY MABEL 0	For Butler Bros. Shoulder head. 1898. 14" $150.00 25" $325.00
MABEL 3/0 GERMANY	For Butler Bros. Shoulder head. 1898. 12" $145.00 26" $345.00
GERMANY MABEL 3/0	For Butler Bros. Shoulder head. 1898. 15" $165.00 30" $425.00
MABEL 5 GERMANY ARMAND MARSEILLE	Turned shoulder head. Also socket head. 1898. 20" $250.00 24" $325.00
MABEL 6/0 GERMANY ARMAND MARSEILLE	Turned shoulder head. 1898. 18" $200.00
GERMANY MABEL CELEBRATE 7	For George Borgfeldt in 1898. Turned shoulder head. 22" $275.00
MABEL 8/0 15½" $185.00	For Butler Bros. Brown shoulder head.
MABEL 9/0x GERMANY	For Butler Bros. 1898. Shoulder head. 17" $200.00
MABEL 10/0 DEP GERMANY	For Butler Bros. 1898. Shoulder head. 17" $200.00
GERMANY MABEL 13/0	Shoulder head. 1898. 15" $165.00
GERMANY MABEL 14/0	Shoulder head. 1898. 12" $145.00
MABEL GERMANY 131	Shoulder head. 1898. 13" $150.00 17" $200.00

GERMANY MELITTA A. 8 M.	Socket head. 1912. Baby. 18″ $345.00
MELITTA GERMANY 12	Baby. Toddler. Open mouth. 1912. 22″ $425.00
MAJESTIC A. 10 M. MADE IN GERMANY	1892. Socket head. (Majestic also made by E.U. Steiner) 14″ $150.00 26″ $345.00
A.M. GERMANY MY COMPANION	For Louis Wolfe & Co. in 1911. (Made by other companies also.) 18″ $265.00
MY DEARIE D.R.G.M. 246/1 A. 2 M.	For Geo. Borgfelt. 1908. Socket head. 14″ $150.00 23″ $275.00 27″ $315.00
A.M. GERMANY MY PLAYMATE	For Geo. Borgfeldt. 1903. Closed mouth, cloth body. 18″ $325.00
A.M. GERMANY MY PLAYMATE (BODY)	Closed dome and mouth. 18″ $325.00
A 253 M 11/0 NOBBIKID U.S. PAT. GERMANY	Closed mouth googly. Socket head. 1915. 7½″ $650.00 12″ $950.00
QUEEN LOUISE GERMANY	Socket head. 15″ $165.00
QUEEN LOUISE 3 GERMANY	Socket head. 1910. For Louis Wolfe. 24″ $300.00
QUEEN LOUISE GERMANY 5/0	Socket head. 1910. For Louis Wolfe. 13″ $145.00
QUEEN LOUISE & GERMANY	Socket head. 1910 for Louis Wolf. 22″ $275.00
28⁸ QUEEN LOUISE GERMANY 7	For Louis Wolf. 1910. Socket head. 22″ $275.00
QUEEN LOUISE GERMANY 8′	For Louis Wolf. 1910. Socket head. 23″ $285.00
L. 9 QUEEN LOUISE GERMANY	Socket head. 1910 for Louis Wolf. 24″ $300.00
GERMANY QUEEN LOUISE 9	Socket head. 1910 for Louis Wolf. 24″ $300.00
GERMANY QUEEN LOUISE 10	Socket head. 1910 for Louis Wolf. 25″ $325.00
QUEEN LOUISE GERMANY 11	Socket head. 1910 for Louis Wolf. 29″ $425.00
QUEEN LOUISE GERMANY 13	Socket head. 1910 for Louis Wolf. 28″ $400.00

24 QUEEN LOUISE GERMANY I	Socket head. 1910 for Louis Wolf. 15″ $165.00
28 QUEEN LOUISE GERMANY 7	Socket head. 1910. For Louis Wolf. 24″ $300.00
QUEEN 29 LOUISE GERMANY	Socket head. 1910 for Louis Wolf. 12″ $145.00 30″ $450.00
QUEEN LOUISE 100 GERMANY	Socket head. 1910 for Louis Wolf. 12″ $145.00 30″ $450.00
QUEEN LOUISE 315 GERMANY 12	Socket head. 1910. For Louis Wolf. 27″ $375.00
QUEEN LOUISE 2015	Socket head. 1910. For Louis Wolf. 12″ $145.00 30″ $450.00
ROSEBUD A. 3 M	Shoulder head for Max Illfelder in 1902. 18″ $200.00
ROSEBUD A 4/0 M MADE IN GERMANY	Shoulder head. 1902. For Max Illfelder. 15″ $165.00
ROSEBUD 8 GERMANY	Shoulder head. 1902. For Max Illfelder. 26″ $350.00
ROSELAND A. O M.	For Max Illfelder in 1910. 18″ $200.00
ROSELAND A. 2/0x M. MADE IN GERMANY	Shoulder head. 1910. For Max Illfelder. 17″ $200.00
SADIE ARMAND MARSEILLE A. 975 M 2	Socket head. Baby. For Louis Wolfe in 1914. 17″ $325.00
SPECIAL GERMANY	Socket head for Geo. Borgfeldt in 1904. 12″ $125.00 26″ $345.00
ST W&CO. A.M. 5 DEP MADE IN GERMANY	Turned shoulder head. Made for Eu. Steiner, distributed by Louis Wolfe. 24″ $300.00
THE DOLLAR PRINCESS GERMANY A. 3½ M.	Socket head. For Geo. Borgfeldt in 1904. 12″ $125.00 26″ $345.00
THE DOLLAR PRINCESS A. 7 M. GERMANY	Socket head. 1904. For Geo. Borgfeldt. 24″ $300.00 30″ $425.00
A.M.	For the Campbell Soup Co. Googly. 9″ $650.00
A.M. GERMANY	Socket head/Oriental/Voice box. 10″ $365.00
A.M.	Socket head/black. 12″ $365.00
A.M. 8″ $185.00	Puppet baby in blanket.

A.1 M.	Socket head/closed dome. 11″ $195.00
A.M. GERMANY	Like My Dream Baby with open/closed mouth. 10″ $185.00
A.M. GERMANY	Like My Dream Baby with closed mouth. 16″ $235.00
A.11 M. D.R.G.M. GERMANY	Socket head. 26″ $325.00
A.M./X GERMANY 4/0	Socket head/Indian. 11″ $325.00
A.M. 2 DEP.	Socket head. 16″ $185.00
A.M. 6/0 D.E.P. ARMAND MARSEILLE MADE IN GERMANY	Shoulder plate/kid body. Bisque lower arms. 12″ $110.00 26″ $250.00
GERMANY A.M. 7/0	Socket head/Indian. 9½″ $295.00
A. 7/0 M. MADE IN GERMANY	Shoulder head/boy. 14″ $145.00
A. 8/0 M.	Socket head/side glance eyes. 13″ $325.00
A.M. 8/0x	Socket head. 13″ $125.00
GERMANY A.M. 8/0	Shoulder head. 12″ $125.00 26″ $250.00
MADE IN GERMANY A. 12/0 M.	Socket head/intaglio eyes. 7″ $75.00
MADE IN GERMANY A. 12/0 M.	Socket head. Sleep eyes. 7″ $95.00
MADE IN GERMANY A. 3. M.	Closed mouth/intaglio eyes "Bernice" socket head. 1915 18″ $950.00
3.0 A.M. 2/0 DEP ARMAND MARSEILLE MADE IN GERMANY	Shoulder head. 1895. 18″ $225.00
3.0.A.M. 6/0 M	Socket head. 15″ $225.00
GERMANY A. 3/0 M	Socket head. Stick body. 6″ $75.00
3./H/41 K A.M.	Socket head. Baby. 16″ $235.00
A. 3/4 M. MADE IN GERMANY	Shoulder head. 15″ $165.00
G.B. A.M. 3-04	Socket head. Closed dome and mouth for Geo. Borgfeldt. Baby. 18″ $300.00 Girl. 18″ $425.00
G.B. A.M. 3-04	Socket head character with molded top knot. Googly baby. Closed mouth. 12″ $750.00 18″ $1,200.00
A.M. 4 DEP GERMANY	Socket head. Stick body. 18″ $175.00

5/6/3/2K	Baby. Closed dome and mouth. For Geo. Borgfeldt. 16" $235.00
A. 6½ M. GERMANY	Socket head. 17" $170.00
A. 7 M. GERMANY	Socket head. 17" $170.00
G.B. GERMANY A 7/0 M D.R.G.M.	Socket head. 14" $135.00
ARMAND MARSEILLE	Socket head. 18½" $175.00
ARRANBEE 2/0	1924 for Arranbee Doll Co. Flange neck baby. Open or open/closed mouth. 12" $250.00
A.M. 8/0 12" $325.00	Indian.
GERMANY 8/0 A.M.	Socket head. 7½" $85.00
A.M. 8 DEP GERMANY	Shoulder head. 23" $265.00
A.M. D.E.P. GERMANY 9	Shoulder head. 29" $400.00
MADE IN GERMANY C.M. BERGMANN A. 9 M.	Socket head for C.M. Bergmann. 24" $325.00
A.M. 9/0 DEP. GERMANY	Socket head. 1895. 12" $150.00
GERMANY A. 10/0 M.	Socket head. 5" $65.00
M.H. 2/00	Scowling Indian. 1890. For Max Handwerck in 1900. 8" $145.00 12" $185.00
OTTO GANS A 11 M	Socket head. Baby. 26" $550.00
ARMAND MARSEILLE GERMANY A. 11 M.	Socket head. Walker, head turns. 26" $345.00
A. 11 M.	Socket head. 26" $250.00
A. 11/0 M GERMANY D.R.G.M.	Socket head. Painted eye googly. 7½" $425.00
MADE IN GERMANY C.M. BERGMANN A. 12 M.	Socket head. 27" $385.00
A.M. GERMANY 12	Flange neck. 1907. 16" $235.00
GERMANY A. 12/0 M.	Socket head. 7" $85.00
GERMANY A. 12 M.	Socket head. 29" $350.00

A.M. 13 D.E.P. GERMANY	Socket head. 28″ $325.00
A.M. GERMANY 15	Flange neck. 1907. 24″ $485.00
A. 16 M. GERMANY	Socket head. 33″ $575.00
GERMANY A.M. 16/0	Socket head. Closed mouth. Googly. 6″ $495.00
GERMANY A. 16/0 M.	Socket head. 5″ $65.00
A. 17. M.	Socket head. 32″ $500.00
GERMANY A. 19 M.	Socket head. 42″ $1,000.00
20 A.M.D.E.P.	Shoulder head. 21″ $195.00
GERMANY A. 20 M.	Socket head. 44″ $1,100.00
A.M. 20 DEP. 21″ $195.00	Shoulder head.
GERMANY A.M. 20/0	Indian. 1890s. Socket head. Open mouth. 8″ $225.00
A.M. 20/0	All bisque Indian. 6″ $165.00
25 A. 11/0 M. GERMANY D.R.G.M.	Socket head googly. Closed mouth. 1912. 8″ $650.00
MADE IN GERMANY 30x	Shoulder head. Closed mouth. 1898. 24″ $875.00
GERMANY M.H. A. 30 M.	"Missy" Socket head lady for Max Handwerck. Closed mouth. 1913. 24″ $1,800.00
GERMANY M.H. A 30/0 M.	"Missy" for Max Handwerck. Adult lady. Socket head. Closed mouth. Dimples. 1913. 10″ $500.00
A.M. GERMANY 34½	Socket head. Closed dome and mouth 18″ $350.00
A.M. 93 7 DEP. MADE IN GERMANY A.M.	Shoulder head. 28″ $325.00
95 4 DEP. ARMAND MARSEILLE MADE IN GERMANY	Shoulder head. 20″ $195.00
95 IV DEP. ARMAND MARSEILLE MADE IN GERMANY	Shoulder head. 20″ $195.00
A.M. 95 5 DEP.	Shoulder head. 19½″ $195.00
A.M. 95 6 DEP. ARMAND MARSEILLE	Turned shoulder head. 20″ $210.00

A.M. 95 8 DEP. GERMANY-ARMAND MARSEILLE	Shoulder head. 22″ $210.00
GERMANY 99 A. 10 M.	Socket head. Baby 20″ $195.00
100 QUEEN LOUISE	Socket head. 1910. 18½″ $200.00
A.M. 110	Socket head. 12″ $150.00 16″ $195.00 26″ $300.00
A.M. 118	"Fluffy Ruffles" 1908. For Stamstag & Hilder. Socket head. 24″ $265.00
GERMANY 1908	Socket head. Pierced ears. 16″ $195.00
W. DEP. C. 121 A.M. GERMANY 131 A. 6 M.	Socket head. 20″ $250.00
A. 140 M.	Socket head. Painted on shoes and socks. 6″ $165.00
ARMAND MARSEILLE GERMANY 147 A. 8 M.	Socket head. 25″ $275.00
200 A. 3/0 M. D.R.G.M. 213	Socket head. Googly. 11½″ $995.00
200 A. 6/0 M. GERMANY D.R.G.M. 243	Socket head. Googly. 9″ $895.00
210 A 5/0 M. GERMANY	Socket head. Baby. Googly. Painted Eyes. 10½″ $795.00
ARMAND MARSEILLE 210 A 8/O M GERMANY D.R.G.M.	Socket head. Closed mouth googly. Painted eyes. 8½″ $445.00
210 A. 10/0 M. GERMANY D.R.G.M.	Socket head, closed mouth googly. Molded hair, painted eyes. 7″ $300.00
210 A. 11/0 M. GERMANY	Socket head. Painted eyes. Googly. 8″ $425.00 12″ $795.00
210 A. 12/0 M. GERMANY	Socket head. Painted eyes. Googly. 6″ $225.00
210 A. 12 M. GERMANY	Socket head. Painted eyes. Googly. 8″ $425.00 12″ $795.00
212 A. 10/0 M. GERMANY	Socket head. Painted eyes. Googly. 7″ $300.00

217
ARMAND MARSEILLE

GERMANY
A 12 M.

Socket head.
12" $150.00 16" $195.00

222
G.B.
A. 7 M. DEP.

"Little Ann"
Socket head for Geo.
Borgfeldt in 1915.
22" $275.00

GERMANY
225
ARMAND MARSEILLE
A.M.

"Little Mary." Socket head.
Two rows teeth. 1916.
16" $385.00

MADE IN GERMANY
D.R.G.M.

A.M.
231

"Fany" character baby. 1913.
Closed mouth. 20" $2,000.00

FANY 231
D.R.G.M. 248/1
A 2/0 M.

Character baby. Closed mouth.
1913.
9" $435.00

FANY 231
D.R.G.M. 248/1
A. 7 M.

Character baby. 1913.
Closed mouth.
19" $1,900.00

FANY 231
D.R.G.M. 248/1
A. 11 M.

Character baby. 1913.
Closed mouth.
25" $2,900.00

A   Ellar   M
231

Oriental baby.
15" $650.00 19" $995.00

A   Ellar   M

231
GERMANY
2K

Socket head. For Paul
Revere Pottery in 1925.
15" $650.00 17" $895.00

MADE IN GERMANY
BABY PHYLLIS
240 2

Cloth body. For Baby Phyllis
Doll Co. in 1915.
12" $225.00 23" $375.00

MADE IN GERMANY
ARMAND MARSEILLE
A 1 M 232
D.R.G.M.

Socket head. Googly with
glass eyes.
10" $795.00

240
BABY PHYLLIS

Cloth body. For Baby Phyllis
Doll Co. in 1915.
26" $500.00

MADE IN GERMANY
ARMAND MARSEILLE
D.R.G.M. 246 1

Socket head. "My Dearie." 1908.
11½" $110.00 16" $165.00

MADE IN GERMANY
ARMAND MARSEILLE

FLORODORA
D.R.G.M. 246 1

Socket head. "My Dearie" 1908.
11½" $110.00 16" $165.00

D.R.G.M. 248
FANY
A. 7 M.

Character baby. Socket head. 1913.
Closed mouth. 22" $2,500.00

G.B.
GERMANY
A 5/0 M
D.R.G.M. 248/1

Socket head for Geo.
Borgfeldt in 1912. Open
mouth.
10" $110.00 20" $195.00

249
D.R.G.M.
GERMANY
ARMAND MARSEILLE

All bisque googly. Painted
eyes & molded hair.
7" $300.00

G.B. 250
A 2/0 M.
GERMANY

Socket head. Closed mouth.
Molded hair for Geo. Borgfeldt.
9½" $475.00

G. 250 B.
A. 7/0 M.
GERMANY

Molded hair, closed mouth
googly for Geo. Borgfeldt.
Painted eyes. 14" $1,000.00

G.B. 250
A. 1 M.
GERMANY

Socket head. Closed mouth
with molded tongue. Molded
hair. 10½" $795.00

G. 251 B.
A. 1 M.

"Dotty" for Geo. Borgfeldt
in 1913. Shoulder head.
16" $245.00

G. 251 B.
D.R.G.M. 243/1
A. 2/0 M.
GERMANY

Socket head. Open/closed mouth.
Pouty for Geo. Borgfedlt.
12" $350.00 16" $495.00

251
G.B.
GERMANY
A. 2/0 M. 243/1 D.R.G.M.

Socket head. OPEN MOUTH.
Pouty for Geo. Borgfeldt.
12" $150.00 16" $195.00

251
G.B.
GERMANY
A. 2/0 M.
D.R.G.M. 243/1

Socket head. Closed Mouth.
For Geo. Borgfeldt.
12" $350.00 16" $495.00

251
G.B.
GERMANY
A. 3/0 M.
D.R.G.M.
8
24x/1

Socket head. Open/closed
mouth/2 teeth and tongue. Baby.
11" $350.00 15" $495.00

251
G.B.
GERMANY
A. 6/0 M.
D.R.G.M.
248/1

Socket head. Open/closed
mouth/2 teeth and molded tongue.
Baby for Geo. Borgfeldt.
8" $250.00 12" $375.00

251
G.B.
A 9/0 M.
D.R.G.M.
248/6

Socket head baby for Geo.
Borgfeldt. Open/closed mouth/
2 teeth and molded tongue.
8" $450.00

GERMANY
G 252 B
A 3/0 M
D.R.G.M.

Socket head googly. 1915.
"Little Bright Eyes."
For. Geo. Borgfeldt.
9½" $795.00

G.B. 252
GERMANY
A 6/0 M. D.R.G.M.

"Little Bright Eyes." Closed
mouth and dome. Socket head.
Googly for Geo. Borgfeldt.
1915. 10" $795.00

S.B. 252
GERMANY
A. 6/0 M.
D.R.G.M.

Socket head. Painted eyes.
Googly for Scheller & Bautler.
10" $795.00

253
S.B.
GERMANY
A. O M.
D.R.G.M.

Socket head. Googly for
Scheller & Bautler.
8" $650.00

GERMANY
G. 253 B.

"Nobbikid," Socket head.
Googly for Geo. Borgfeldt. 1915.

A 3/0 M
D.R.G.M.

10" $795.00

A. 253 M.
5/0
D.R.G.M. GERMANY

"Nobbikid." Socket head.
Googly for Geo. Borgfeldt. 1915.
16" $1,400.00

G. 253 B.
GERMANY
A. 5/0 M.
D.R.G.M.

"Peero" & "Nobbikid." Googly
for Geo. Borgfeldt. 1915.
10" $795.00

253
A. 5/0 M.
GERMANY
D.R.G.M.

"Peero" and "Nobbikid." Intaglio
eye googly. 1915.
10" $795.00

A. 253 M.
NOBBIKID
U.S. PAT.
GERMANY
5/0

Socket head googly of 1915.
10" $795.00

GERMANY
G. 253 B
A. 6/0 M.
D.R.G.M.

"Peero" and "Nobbikid" for
Geo. Borgfeldt. Googly.
1915.
9" $725.00

G.B. 253
GERMANY
A. 6 M.

"Peero" and "Nobbikid." Socket
head googly for Geo. Borgfeldt.
1915.
8" $650.00 16"

GERMANY
G. 253 B.
A. 8/0 M.
D.R.G.M.

"Peero" and "Nobbikid" googly. 1915
7½ $650.00

A. 253 M. NOBBIKID
REG. U.S. PAT. OFF.
GERMANY
10/0

Googly. 1915.
7½" $650.00

G. 253 B.
GERMANY
A. 11/0 M.

"Peero" and "Nobbikid" googly. 1915.
7½" $650.00

A. 253 M.
U.S. PAT. GERMANY
11/0

"Peero" and "Nobbikid" googly. 1915.
7½" $650.00

A.G. 253 B.M.
A. 11/0 M.
GERMANY

"Peero" and "Nobbikid" googly
for Geo. Borgfeldt. 1915.
7½" $650.00

A. 253 M.
NOBBIKID
U.S. PAT. 11/0
GERMANY

Socket head googly. 1915.
7½" $650.00

254
A. 11/0 M.

Socket head. Googly with
molded hair.
8" $650.00

254 A. 12/0 M.	Socket head. Googly with molded hair. 7″ $600.00
254 A. 13/0 M.	Socket head. Googly with molded hair. 6½″ $600.00
255 A. 9/0 M. D.R.G.M. GERMANY	Socket head. Googly. Painted eyes. 8″ $445.00
A. 255 M. NOBBIKID D.R.G.M. 2 4/0	Socket head. Glass eyes. Googly. 1915. 9½″ $795.00
A. 255 M. NOBBIKID D.R.G.M. 2 8/0	Socket head. Glass eyes. Googly. 1915. 7½″ $650.00
255 D.R.G.M. A. 11/0 M. GERMANY	Socket head. Painted hair. Googly. 7½″ $650.00
255 D.R.G.M. 2 A. 11/o M. GERMANY	Socket head. Intaglio eye. Googly. 7½″ $350.00
MADE IN GERMANY ARMAND MARSEILLE 256	"Little Jane" for E. Maar & Sohn of Monchroden in 1915. 18″ $200.00
A. 3/0 M. MAAR	
257 GERMANY A. 7 M.	Socket head baby. 1914. 22″ $450.00
A. 13/0 M. 258 D.R.G.M. GERMANY	"Cheer Ups" for Geo. Borgfeldt in 1916. Intalio eye googly. 7″ $300.00
MADE IN GERMANY A. 0 M. 259 D.R.G.M.	"First Steps." 1919. Socket head. Baby/toddler. 18″ $350.00
A. 5/0x M. D.R.G.M. GERMANY	Socket head baby/toddler. 16″ $235.00
A. 7/0x M. D.R.G.M. 259 MADE IN GERMANY	"First Steps" toddler for Louis Amberg in 1919. 13½″ $285.00
A. 15 M. D.R.G.M. 259	"First Steps" Socket head baby. 1919. 27″ $575.00
A.M. 264-1 A.G.M.	"Friedel" for E.W. Matthes. Socket head. 1916. 24″ $285.00
265 D.R.G.M. A. 11/0 M. GERMANY	Socket head character baby. 8″ $145.00
266	No information
D.R.G.M. 276/1 A. 11 M. GERMANY	Socket head. 26″ $300.00

288 QUEEN LOUISE GERMANY	1910 socket head. 22″ $275.00
ARMAND MARSEILLE 300 A 13/0 M. GERMANY	Socket head. 8″ $115.00
ARMAND MARSEILLE 300n GERMANY A. 0½ M.	"May Queen" socket head. 1910 for Butler Bros. 15½″ $185.00
A.M. 301	Socket head. 16″ $185.00
GERMANY 310 A 8/0 M.	"Just Me" Socket head. 1928 googly. 8″ $600.00 10″ $900.00 12″ $1,100.00
GERMANY 310 A 8/0 M.	"Just Me." Socket head googly with painted bisque. 8″ $250.00 10″ $395.00 12″ $495.00
A.M. 310 GERMANY	Socket head with "dolly" face, and not "Just Me" googly. 10″ $900.00 16″ $1,500.00
QUEEN LOUISE 315 GERMANY 12	Socket head. 27″ $375.00
320 A.M. O DEP. GERMANY	Socket head. Closed mouth and painted eyes to side googly. Molded hair. 8″ $395.00
320 A.M. 2½ R.E.P. MADE IN GERMANY	"Marguerite" for Hamberger 1902. Socket head. Closed mouth. 17½″ $175.00
320 A. 4/0 M. GERMANY	Indian. Socket head with molded "Mohawk" hair style. 15″ $650.00
320 A. 4/0 M. GERMANY	Socket head. Closed mouth. Painted eye googly. 6½″ $245.00 11″ $795.00
320 A. 11/0 GERMANY 320½	Socket head. Painted eye googly. With or without one strip of painted hair. (Indian or White) 7½ $325.00
320 A.M. 12/0 Dep.	Socket head, closed mouth, googly painted eyes to side Molded hair. 6½″ $245.00
320 A.M. 13/0 Dep. A.M. 322 11 GERMANY	Socket head. Closed mouth, painted eye googly. 6½″ $245.00
A.M. 322 11 GERMANY	Socket head. Closed dome and mouth. One strip molded hair. Googly. 10″ $695.00 15″ $1,000.00
ARMAND MARSEILLE GERMANY 322 A. 4/0 M.	Socket head. Closed mouth. Googly 10″ $695.00 15″ $1,000.00

GERMANY 322 A. 6/0 M.	Closed mouth. Painted eye baby. Googly. 10″ $695.00
322 A. 11/OM. GERMANY	Socket head. Closed dome and mouth. Painted eye googly. 7½″ $325.00 10″ $695.00
GERMANY 323 A. 3/0 M.	Socket head. Googly. 11″ $650.00 13″ $795.00
GERMANY 323 A. 4/0 M.	Socket head. Googly. 11″ $650.00 14″ $850.00
A.M. 323 6/0 GERMANY	Socket head. Googly. 10″ $600.00 16″ $1,000.00
323 A. 6/0 M. GERMANY	Socket head. Googly. 10″ $600.00 13″ $795.00
323 A. 6/0 M. GERMANY	Socket head. Googly. 7½″ $500.00 10″ $600.00
323 A. 8/0 M.	Socket head. Googly. 9″ $550.00 15″ $950.00
ARMAND MARSEILLE GERMANY 323 A. 8/0 M.	Socket head. Googly. 9″ $550.00 13″ $795.00
GERMANY 323 A. 11/0 M.	Socket head. Googly. 7½″ $500.00
GERMANY 323 A. 12/0 M.	Socket head. Googly. 7″ $450.00
GERMANY 323 A. 14/0 M.	Socket head. Googly. 6″ $325.00
A.M. 324	Socket head. Googly. 7″ $300.00 12″ $795.00
ARMAND MARSEILLE GERMANY 324 A. 10/0 M.	Painted eye googly. 7″ $300.00
A. M. 325 GERMANY	"Glad Baby." Socket head. Googly. 1925. 12″ $750.00
325 A. 6/0 M. GERMANY	Socket head. Googly baby. "Glad Baby." 1925. 10″ $650.00 14″ $1,000.00
A.M. 326 GERMANY	"Rosie Baby." Socket head. Two lower molded teeth. 1912. 12″ $250.00 20″ $435.00
A.M. G. 326 B. GERMANY	"Prize Baby" for Geo. Borgfeldt. 1912. Socket head. Baby. 12″ $250.00 20″ $435.00

GERMANY
G. 326 B.
A.O.M.
D.R.G.M. 250

"1st Prize Baby" Socket head
baby for Geo. Borgfeldt.
12" $250.00 20" $435.00

GERMANY
G. 326 B.
A. 1 M.
D.R.G.M. 259

"Prize Baby"
Socket head baby. 1912.
12" $250.00 14" $285.00

GERMANY
G.326 B.
A. 2/0 M.
D.R.G.M. 2

"Little Sister." Socket head
baby for Geo. Borgfeldt.
12" $250.00 20" $435.00

GERMANY
G. 326 B.
D.R.G.M. 2
A. 2/0 M.

"Little Sister" Socket head.
Baby.
10" $195.00 16" $325.00

A.M.
G. 327B
GERMANY

Socket head. Baby.
12" $250.00 15" $325.00

A.M.
G. 327B.
D.R.G.M.

"Jason" for Geo. Borgfeldt.
1914. Socket head. Fur hair.
Baby. 12" $250.00 18" $400.00

G. 327 B.
A.M.
D.R.G.M.

"Jason." Socket head baby for Geo.
Borgfeldt. Fur hair. 1914.
10" $175.00 20" $425.00

GERMANY
G. 327 B.
D.R.G.M. 259
A. 1 M.

"Jason." Socket head baby for
Geo. Borgfeldt. Fur hair. 1914.
13" $265.00 20" $425.00

GERMANY
G. 327 B.
D.R.G.M. 259
A. 2/0 M.

"Jason." Socket head baby for
Geo. Borgfeldt. Fur hair. 1914.
14" $275.00 16" $325.00

GERMANY
G. 327 B.
D.R.G.M. 259
A. 3 M.

"Jason." Socket head
baby for Geo. Børgfeldt. Fur
hair. 1914.
14" $275.00 20" $425.00

GERMANY
G. 327 B.
D.R.G.M. 259
A 5 M

"Jason." Socket head baby for
Geo. Borgfeldt. 1914. 18" $400.00

GERMANY
G. 327 B.
D.R.G.M. 259
A. 6/0 M.

"Jason." Socket head baby for
Geo. Borgfeldt. 1914.
12" $250.00 16" $325.00

G.B.
327
A. 8. M.
D.R.G.M. 259

"Jason." Socket head baby for
Geo. Borgfeldt. 1914.
17" $350.00

GERMANY
G. 327 B.
D.R.G.M. 259
A. 8.M.

"Jason." Socket head used as
girl.1914.
17" $350.00

GERMANY
G. 327 B.
D.R.G.M. 259
A. 10 M.

"Jason." socket head used for
girl. 1914.
20" $400.00

GERMANY
G. 327 B.
D.R.G.M. 259
A 11 M.

"Jason."
Socket head. Baby. 1914.
20″ $400.00

GERMANY
G. 327 B.
A 11 M.

"Jason." Socket head baby for
Geo. Borgfeldt. 1914.
20″ $400.00

GERMANY
G. 327 B.
D.R.G.M. 259
A 12/0 M.

"Jason." Socket head. 1914.
16½″ $325.00

GERMANY
G. 327 B.
A. 12 M.

"Jason." Socket head baby for
Geo. Borgfeldt. 1914.
22″ $465.00

G. 327 B.
GERMANY
A. 13 M.

"Jason." Socket head baby for
George Borgfeldt. 1914.
25″ $500.00

G. 327 B.
D.R.G.M. 259
A. 14 M.

"Jason." Socket head baby for
Geo. Borgfeldt. 1914.
26″ $550.00

G. 328 B.
A.M.
GERMANY

"Baby Bobby" for Geo. Borgfeldt
in 1922. Socket head. Closed
dome baby.
14″ $285.00

GERMANY
G. 328 B.
D.R.G.M. 267/1
A. 4 M.

"Baby Bobby" for Geo. Borgfeldt
in 1922.
14″ $285.00 20″ $425.00

GERMANY
G. 328 B
D.R.G.M. 267/1
A 5 M.

"Baby Bobby" for Geo. Borgfeldt
in 1922.
15″ $300.00 22″ $465.00

G. 328 B.
A 6 M
GERMANY

"Baby Bobby" for Geo. Borgfeldt
in 1922. Painted hair.
16″ $325.00

GERMANY
G. 329 B
A.O.M.
D.R.G.M. 267/1

"Betsy Baby" for Geo. Borgfeldt
in 1922. Socket head baby.
11½″ $250.00 14″ $275.00

G.329B.
A.M.
D.R.G.M. 267/1
GERMANY

"Betsy Baby" for Geo. Borgfeldt
in 1922. Socket head baby.
16″ $325.00 20″ $425.00

GERMANY
G. 329 B.
A. 1 M.
D.R.G.M. 267/1

"Baby Betsy" for Geo. Borgfeldt
in 1922. 16″ $325.00 18″ $400.00

G. GERMANY B.
329
A. 1 M.
D.R.G.M. 267/1

"Betsy Baby." Socket head baby
in 1922 for Geo. Borgfeldt.
16″ $325.00 24″ $500.00

G. 329 B.
A. 1 M.
D.R.G.M. 267/1

"Betsy Baby" baby for Geo.
Borgfeldt in 1922.
12″ $250.00 15″ $300.00

GERMANY
G.329 B.
A. 2/0 M.

Socket head girl.
9″ $165.00 16″ $350.00

GERMANY
G. 329 B.
A. 5 M.
D.R.G.M. 267/1

"Betsy Baby" for Geo. Borgfeldt
in 1922. Socket head baby.
15" $300.00 18" $400.00

GERMANY
G.329B.
A. 7 M.
D.R.G.M. 267/1

"Betsy Baby" for Geo. Borgfeldt
in 1922. Socket head baby.
18" $400.00 25" $550.00

GERMANY
G. 329 B.
A. 8 M.
D.R.G.M. 267/1

"Betsy Baby" for Geo. Borgfeldt
in 1922. Socket head baby.
22" $465.00 27" $600.00

329/13
A.M.

All bisque. Socket head.
7½" $135.00

A.M. 341
"Polly" in red
script stamp.

Bisque head. Wires through arms.
press stomach and hands clap and
she cries "mama." 1923.
12" $285.00 16" $385.00

A.M.
341
GERMANY

"My Dream Baby" for Arranbee
Doll Co. Flange neck. Closed
mouth. 1924.
7" $150.00 14" $300.00

A.M.
341
GERMANY

Brown "My Dream Baby" for
Arranbee. 1924. Closed mouth.
6½" $200.00 16" $350.00

341
A.M.
GERMANY

Flange and socket. "My Dream
Baby." Closed mouth.
White: 7" $150.00 16" $300.00
Black: 7" $250.00 16" $450.00

A.M.
GERMANY
341/0

Flange or socket. "My Dream
Baby." Closed mouth.
White: 9" $165.00 13" $285.00
Black: 9" $235.00 13" $385.00

A.M.
GERMANY
341/Ok

Socket head. "My Dream Baby."
Closed mouth.
7" $150.00 16" $325.00

A.M.
GERMANY
341/1k

Flange. "My Dream Baby."
Closed mouth.
7" $150.00 20" $425.00

A.M.
GERMANY
341-2

Flange. "My Dream Baby."
Closed mouth.
12" $250.00 18" $385.00

A.M.
GERMANY
341 21 2k

Socket head. "My Dream Baby."
Closed mouth.
White: 9" $165.00 16" $285.00
Black: 9" $235.00 16" $385.00

A.M.
GERMANY
341/2k

Socket and flange. "My Dream
Baby." Closed mouth.
7" $150.00 20" $425.00

A.M.
GERMANY
341/2/Ok

Hawaiian. Flange or socket.
"My Dream Baby." Closed mouth.
9" $235.00 15" $385.00

341
3 A.M.
GERMANY

Flange. "My Dream Baby."
6½" $150.00 13" $275.00

A.M.
341/3
GERMANY

Socket and flange. "My Dream
Baby." Closed mouth.
White: 8" $165.00 15" $300.00
Black: 8" $235.00 15" $350.00

A.M. GERMANY 341/3k	Flange. "My Dream Baby." Closed mouth. White: 7" $150.00 14" $300.00 Black: 7" $235.00 14" $350.00
A.M. GERMANY 341 3½k	Socket head. "My Dream Baby." Closed mouth. 15" $300.00
A.M. GERMANY 341/4	Flange. "My Dream Baby." Closed mouth. 15" $300.00
A.M. GERMANY 341/4k	Flange. "My Dream Baby." Closed mouth. Black: 15" $350.00
A.M. GERMANY 341/4/0	Flange. "My Dream Baby." Closed mouth. 8" $165.00
A.M. GERMANY 341/5k	Flange. "My Dream Baby." Closed mouth. White: 15" $300.00 Black: 15" $350.00
A.M. GERMANY 341 6/0	Flange. "My Dream Baby." Closed mouth. 7" $150.00
A.M. GERMANY 341/6	Flange. "My Dream Baby." Closed mouth. 16½" $350.00
A.M. GERMANY 341 7	Flange. "My Dream Baby." Closed mouth. 18" $385.00
A.M. GERMANY 341 8/0	Flange. "My Dream Baby." Closed mouth. 6½" $150.00
A.M. GERMANY 341-8	Flange. "My Dream Baby." Closed mouth. 18" $385.00
A.M. GERMANY 341 8k	Flange. "My Dream Baby." Closed mouth. White: 22" $425.00 Black: 22" $525.00
A.M. GERMANY 341/10	Flange. "My Dream Baby." Closed mouth. 21" $425.00
A.M. GERMANY 341 11k	Socket. "My Dream Baby." Closed mouth. White: 21" $425.00 Black: 21" $525.00
A.M. GERMANY 341 12	Flange. "My Dream Baby." 17" $365.00
A.M. GERMANY 341 14/0	Flange. "My Dream Baby." Closed mouth. White: 6" $150.00 Black: 6" $200.00
A.M. GERMANY 15/0x	Flange and socket. "My Dream Baby." Closed mouth. White: 6½" $150.00 Black: 6½" $200.00
A.M. GERMANY 341/150x K	Flange and socket. "My Dream Baby." Closed mouth. Some marked Kiddijoy. 7" $150.00 18" $385.00

A.M. GERMANY 342 1	Flange. Pillow puppet. Open/ closed with two lower teeth. Molded hair. 11″ $285.00 18″ $425.00
342/4	Socket in shoulder plate. Cloth body. Open/closed with teeth. Press stomach and plays music "Home Sweet Home" 16″ $385.00
A.M. GERMANY 347/0	"Bumble Puppy" for Max Illfelder. 1909. Socket. 16″ $325.00
A. 4 M. GERMANY 347	"Bumble Puppy." Socket head. 1909. 16″ $325.00
A.M. GERMANY 347/4	"Bumble Puppy" baby. Socket head. 1909. 12″ $250.00 16″ $325.00
A.M. GERMANY 350/9	Flange. Molded hair. Open mouth baby. 7″ $150.00 18″ $385.00
A.M. 351 O.K.	"My Dream Baby." Socket. Open mouth. Painted eyes. 7″ $150.00 14″ $300.00
351 A.M. O.K. GERMANY	Socket. "My Dream Baby." Open mouth. 12″ $250.00 16″ $350.00
A.M. 351 O.K. GERMANY	"My Dream Baby" Open mouth. 7″ $150.00 18″ $385.00
A.M. GERMANY 351 1 k	Socket. "My Dream Baby." Open mouth. White: 10″ $175.00 16″ $300.00 Black: 10″ $235.00 16″ $350.00
A.M. 351/2k GERMANY	Socket. "My Dream Baby." Open mouth. White: 14″ $300.00 20″ $425.00 Black: 14 $350.00 20″ $500.00
A.M. GERMANY 351 2/0k	Flange. "My Dream Baby." Open mouth. White: 10″ $175.00 Black: 10″ $235.00
A.M. GERMANY 351 2/0	Flange. "My Dream Baby." Open mouth. 9½″ $175.00
A.M. GERMANY 351 2½	Flange. "My Dream Baby." Open mouth. 12″ $250.00
A.M. GERMANY 2½k	Socket. "My Dream Baby." Open mouth. Black: 12″ $295.00
A.M. GERMANY 351 3k	Socket. "My Dream Baby." Open mouth. White: 14″ $300.00 Black: 14″ $350.00
A.M. 351 3½ K	Flange. "My Dream Baby." Open mouth. White: 15″ $315.00 Black: 15″ $365.00
A.M. GERMANY 351/4k	Socket and flange. "My Dream Baby." Open mouth. White: 22″ $425.00 Black: 22″ $525.00

A.M.
GERMANY
351 4/0

Flange. "My Dream Baby."
Open mouth.
9" $165.00

A.M.
351 5/0
GERMANY

Flange or socket. "My Dream
Baby." Open mouth.
White: 5½" $135.00 Black: 5½" $185.00

A.M.
GERMANY
351 L 5K

Socket or flange. "My Dream
Baby." Open mouth.
White: 24" $485.00 Black: 24" $575.00

A.M.
GERMANY
351 6/0

Socket or flange. "My Dream
Baby." Open mouth.
White: 8" $165.00 Black: 8" $235.00

A.M.
351/6k

Socket. "My Dream Baby." Open
mouth.
White: 25" $500.00 Black: 25" $600.00

A.M.
GERMANY
351/6

Flange and socket. "My Dream
Baby." Open mouth.
26" $550.00

A.M.
GERMANY
351 7

Flange. "My Dream Baby."
Open mouth.
20" $425.00

A.M. GERMANY
351/7k
Black: 20" $500.00

Socket. "My Dream Baby."
Open mouth.

A.M.
GERMANY
351 7½k

Socket. "My Dream Baby."
Open mouth.
Black: 21" $500.00

A.M.
GERMANY
351/8k

Socket. "My Dream Baby."
Open mouth.
Black: 24" $595.00

A.M.
351 10/0
GERMANY

"Wee One." Colored with rubber
body. 1922.
6½" $165.00 14" $265.00

A.M.
GERMANY
351 10/0

"Wee One" with rubber body.
7" $165.00 15" $275.00

A.M.
GERMANY
351 13k

Socket. "My Dream Baby."
Open mouth.
Black: 25" $625.00

A.M.
GERMANY
351 13/0k

Socket. "My Dream Baby."
Open mouth.
Black: 9½" $235.00

A.M.
GERMANY
351. 14.k

Socket. "My Dream Baby."
Open mouth.
26" $650.00

A.M.
GERMANY
351 14/Ok

Socket. "My Dream Baby."
Open mouth.
White: 7½" $150.00 Black: 7½" $200.00

A.M.
GERMANY
351 17/0

Socket and flange. "My Dream Baby."
Open mouth.
6" $135.00

A.M.
351 20/0k

Socket. "My Dream Baby."
Open mouth.
Black: 5" $145.00

A.M.
GERMANY
351

All papier mache "My Dream
Baby." 12" $85.00 18" $250.00

A.M.
352
GERMANY

"Baby Love." Socket Head. 1914.
12" $200.00 18" $325.00

A.M.
352
GERMANY 11

"Baby Love."
Socket/shoulder plate. Baby.
Cloth body/voice box. 1914.
21" $365.00

A.M.
352
1k

"Baby Love" Socket. 1914.
White: 9" $160.00 20" $345.00
Black: 9" $200.00 20" $425.00

A.M.
GERMANY
352/8

Flange. "Baby Love." 1914.
23" $425.00

A.M.
GERMANY
352 10

Flange. "Baby Love." 1914.
19" $345.00

A.M.
GERMANY
353

Oriental. Socket.
9" $450.00 14" $650.00

A.M. 353
O.K.
GERMANY

Oriental. Socket head.
6½" $285.00 14" $650.00

A.M.
GERMANY
353 2½k

Oriental. Socket. Closed
mouth and dome.
12" $550.00 18" $925.00

A.M.
353 3k
GERMANY

Oriental. Socket. Closed
mouth and dome.
8" $400.00 13" $600.00

A.M.
GERMANY
353 4k

Oriental. Socket. Closed
mouth and dome.
16" $675.00 21" $1,050.00

A.M.
353
5/0

Socket head googly.
6½" $285.00 9" $450.00

A.M.
GERMANY
353 12/0k

Oriental. Socket. Closed
mouth and dome.
6½" $285.00 8" $400.00

A.M.
362
GERMANY

"Teenie Weenie."
Socket. Baby with closed
dome. 1922. For Geo. Borgfeldt.
White: 12" $250.00 Black: 12" $295.00

362
A.M. 3k
GERMANY

"Teenie Weenie."
Socket head. Baby. Closed
dome. 1922. For Geo. Borgfeldt.
White: 15" $300.00 Black: 15" $385.00

362
GERMANY
A. 5 M.
D.R.G.M. 267/1

"Teenie Weenie."
Socket head. Cryer. Baby. 1922.
For Geo. Borgfeldt.
White: 20" $425.00 Black: 20" $500.00

370
A.M. 0½ DEP.
ARMAND MARSEILLE
MADE IN GERMANY

Shoulder head.
19" $185.00

370
A.M. 1 DEP.

Shoulder head.
20" $195.00

GERMANY
370

Shoulder head.
12" $120.00

30″ $400.00
A.M. DEP.

370   "Banker's Daughter" for
A.M. 2 DEP.   Butler Bros. 1893.
                                       21″ $200.00

370   Shoulder head.
A.M. 2/0x DEP.   19½″ $195.00
MADE IN GERMANY

370   Shoulder head. Fur eyebrows.
A.M. 2/0x DEP.   19½″ $195.00
ARMAND MARSEILLE
MADE IN GERMANY

370   Shoulder head.
A.M. 2/0 DEP.   19½″ $195.00
MADE IN GERMANY

370   Shoulder head.
A.M. 2/0 DEP.   19½″ $195.00
ARMAND MARSEILLE
MADE IN GERMANY

ARMAND MARSEILLE   Shoulder head.
370   19½″ $195.00
A.M. 2/0 DEP.

ARMAND MARSEILLE   Shoulder head.
MADE IN GERMANY   21″ $200.00 370
A.M. 2½ DEP.

370   Brown tones. Shoulder head.
A.M. 2½ DEP.   22″ $295.00
MADE IN GERMANY

370   Shoulder head. Cotton body.
A.M. 3 DEP.   1910.
MADE IN GERMANY   22″ $210.00

370   Shoulder head. 1905.
A.M. 3 DEP.   24″ $225.00
MADE IN GERMANY

370   Shoulder head. Fur eyebrows.
A.M. 3 DEP.   22½″ $225.00
MADE IN GERMANY

370   Shoulder head.
A.M.   15″ $150.00
3/0
DEP.

370   Shoulder head. Fur eyebrows.
D.R.G.M. 4830 37430   22″ $225.00
A.M. 3 DEP.

ARMAND MARSEILLE   "Our Ann" for P.D.G. Co.
370 12   Shoulder head. 1900.
A.M. 3/0 DEP.   16″ $165.00

370   Shoulder head.
A.M. 4 DEP.   23″ $215.00
ARMAND MARSEILLE
MADE IN GERMANY

A. 4 M. 370 DEP.   Shoulder head.
ARMAND MARSEILLE   23″ $215.00
MADE IN GERMANY

370   Shoulder head.
A.M. 4/0x DEP.   16½″ $170.00
MADE IN GERMANY

370
A.M. 4/0x DEP.

Shoulder head.
16½" $170.00

370
A.M. 4/0 DEP.
MADE IN GERMANY

Shoulder head.
15" $150.00 17" $170.00

370
A. 4/0 M.

Shoulder head.
17" $170.00

MADE IN GERMANY
370
A.M. 5/0x

Shoulder head. Closed mouth.
15½" $155.00

370
A.M. 5/0x
MADE IN GERMANY

Shoulder head.
15½" $155.00

370
A.M. 5/0 DEP.
MADE IN GERMANY
A.M.

Shoulder head.
16" $165.00

370
A.M. 5/0 DEP.
MADE IN GERMANY

Shoulder head.
17" $170.00

370
A.M. 5/0 DEP.
ARMAND MARSEILLE
MADE IN GERMANY

Shoulder head.
16" $165.00

370
A.M. 5 DEP.

Shoulder head.
24" $225.00

370
D.R.G.M. 377439
A.M. 5 DEP.
MADE IN GERMANY

Shoulder head. Fur eyebrows.
26" $285.00

ARMAND MARSEILLE
GERMANY 3
370
A. 6/0 M.

"Gold Coast Girl" Shoulder
head. 1905.
15" $150.00

ARMAND MARSEILLE
GERMANY
370-3
A. 6/0 M.

"Gold Coast Girl" Shoulder head.
1905.
15" $150.00

370
A.M. 6/0 DEP.

Shoulder head.
15" $150.00

370
A.M. 7 DEP.

Shoulder head.
26" $250.00

370
A.M. 7/0 DEP.

Shoulder head. Body Sticker:
"Florodora"
14" $150.00

ARMAND MARSEILLE
MADE IN GERMANY
370n
A.M. 7/0 DEP.

Shoulder head.
12" $120.00

ARMAND MARSEILLE
GERMANY
370
A.M. 7/0x DEP.

Shoulder head. 1905.
13½" $135.00

370
A.M. 8/0x DEP.
ARMAND MARSEILLE
MADE IN GERMANY

Shoulder head.
14" $135.00

370 A.M. 9 DEP. ARMAND MARSEILLE	"My Princess" 1905. Shoulder head. 29" $350.00
GERMANY 370 A. 9/0 M.	Shoulder head. 12" $120.00
370 A.M. 10/0 DEP. ARMAND MARSEILLE MADE IN GERMANY	Shoulder head. 12" $120.00
370 A.M. 10 DEP. ARMAND MARSEILLE MADE IN GERMANY	Shoulder head. Fur eyebrows. 32" $550.00
A.M. D.R.G.M. 2 DEP. 370 374830 31 GERMANY	"Miss Myrtle" for Geo. Borgfeldt. 1899. Shoulder head. Fur eyebrows. 21" $250.00
370 4830/3 370 D.R.G.M. 3 04830 34631 A.M. 2/0x DEP. MADE IN GERMANY	"Miss Myrtle" for Geo. Borgfeldt. 1899. Fur eyebrows. Shoulder head. 19½" $225.00
A.M. D.R.G.M. 2 DEP. 370 3748307 31 GERMANY	"Miss Myrtle" for Geo. Borgfeldt. Shoulder head. Fur eyebrows. 21" $250.00
GERMANY Kiddiejoy 372	Shoulder head. Molded hair. 1926. 9" $285.00 18" $625.00
A. 1 M. 372 A.M.	"Banner Kid Dolls" 1894. ½ kid and ½ muslin. 15" $185.00 21" $250.00
372 A.M. 11/0 M GERMANY	Socket head googly with molded hair. 7" $300.00
375 GERMANY A.M. Kiddiejoy	For Hitz, Jacobs & Kassler in 1918. Molded hair. 12" $300.00 22" $825.00
GERMANY KIDDIEJOY 375 6	Molded hair girl. Glass eyes. Closed mouth. 20" $750.00
376 A.M.	Shoulder head. 20" $195.00
D.R.G.M. 377 139 Florodora	Shoulder head. 32" $500.00
A.M. 380 2½ DEP. GERMANY	"Little Sweetheart" for Max Illfelder in 1902. Shoulder head. 17½" $175.00
380 A.M. 2½ DEP. GERMANY	"Little Sweetheart" for Max Illfelder in 1902. Shoulder head. 17½" $175.00
387	No information
A.M. GERMANY 390	Socket head. Papier mache. 11" $115.00

A.M.
GERMANY
390

Socket head.
12" $120.00 16" $165.00

ARMAND MARSEILLE
MADE IN GERMANY
390
A O M.

Brown socket head.
15" $225.00

MADE IN GERMANY
390
A. 0½ M.

Socket head.
15½" $165.00

A.M. 390
A. 0½ M.
MADE IN GERMANY

Socket head.
15½" $165.00 19" $185.00

ARMAND MARSEILLE
390
A. 1 M.

Socket head.
16" $165.00

ARMAND MARSEILLE
GERMANY
390n
A. 1 M.

Brown socket head.
16" $225.00

MADE IN GERMANY
ARMAND MARSEILLE
390n
D.R.G.M. 246/1
A. 1 M.

Socket head. "My Dearie"
1908 to 1922.
18" $225.00

MADE IN GERMANY
ARMAND MARSEILLE
390n
D.R.G.M. 246/1
A. 1½ M.

"My Dearie" socket head for
Geo. Borgfeldt. 1908 to 1922.
16½" $215.00

ARMAND MARSEILLE
390n GERMANY
A. 2 M.

"Patrice" Socket head.
18" $250.00

390
A. 2 M.

Socket head. Talker ("Mama").
17" $225.00

390
A. 2 M.
3

Socket head. Spring joints.
17" $185.00

ARMAND MARSEILLE
390
A. 2½ M.

Socket head.
17½" $175.00

MADE IN GERMANY
ARMAND MARSEILLE
390 D.R.G.M. 246/1
A 2/0 M.

Socket head. "My Dearie" for
Geo. Borgfeldt in 1908 to
1922.
12" $120.00 23" $215.00

ARMAND MARSEILLE
D.R.G.M. 246/1
390 A. 2/0x M.

"My Dearie" for Geo. Borgfeldt
in 1908 to 1922. Socket head.
12½" $135.00 26" $250.00

MADE IN GERMANY
ARMAND MARSEILLE
390
A 3/0 M.

Socket head.
White: 13" $130.00 Black: 13" $180.00

MADE IN GERMANY
390
A. 3/0s M.

Gibson Girl. Socket head.
11½" $150.00 19" $200.00

390
A.M. 246/1
A. 3 M.

"My Dearie" for Geo. Borgfeldt
in 1908 to 1922. Socket head.
22" $210.00 30" $400.00

A.M.
390
D.R.G.M. 240
3½

"Dotty" for Geo. Borgfeldt in
1913. Socket head.
18½" $185.00 28" $325.00

MADE IN GERMANY
A.M. 390

"My Dearie" socket head for
Geo. Borgfeldt. 1908 to 1922.

D.R.G.M. 246/1
A. 3½ M.

18½" $185.00 30" $400.00

MADE IN GERMANY
390 A. 4 M.

Socket head.
19" $185.00

390n
D.R.G.M.
A. 4 M.
MADE IN GERMANY

Socket head.
19" $185.00 25" $235.00

ARMAND MARSEILLE
390
D.R.G.M. 346/1
A. 6/0 M.

Socket head.
11" $115.00 17" $170.00

MADE IN GERMANY
ARMAND MARSEILLE
390
D.R.G.M. 346/z
A. 4 M.

Closed mouth. Socket head.
19" $850.00

MADE IN GERMANY
ARMAND MARSEILLE
390n
A. 4/0 M.

Socket head.
11" $165.00

390
A. 4 M.
GERMANY

Socket head.
19" $185.00 29" $350.00

ARMAND MARSEILLE
GERMANY
390n

Socket head.
12" $170.00 30" $450.00

ARMAND MARSEILLE
GERMANY
390n
A. 4/0x M.

"Louisa." 1915. Socket
head.
11" $165.00

390
A. 5/0 M.
GERMANY

Socket head.
10½" $110.00 19" $185.00

ARMAND MARSEILLE
GERMANY
390
A. 5 M.

Socket head.
20" $195.00 30" $400.00

MADE IN GERMANY
ARMAND MARSEILLE
390
D.R.G.M. 245/1
A. 5 M.

"Pretty Peggy" socket head
for Geo. Borgfeldt in 1909.
20" $195.00 28" $325.00

MADE IN GERMANY
ARMAND MARSEILLE
390n
D.R.G.M. 246/1
A. 5½ M.

"My Dearie" for Geo. Borgfeldt
in 1908 to 1922. Socket head.
16" $210.00 20½" $250.00

ARMAND MARSEILLE
390
A. 6 M.

Socket head.
21" $200.00 27" $300.00

MADE IN GERMANY
390
A. 6 M.

Socket head.
11½" $120.00 21" $200.00

ARMAND MARSEILLE
390n
A 6/0 M.
D.R.G.M. 216/1
A. 6/0 M.

"Mimi" for Geo. Borgfeldt
in 1922.
10" $145.00 16" $210.00 20" $245.00

MADE IN GERMANY
390n
D.R.G.M. 246/1
A. 6 M.

"My Dearie" for Geo. Borgfeldt
in 1908 to 1922. Socket head.
22" $265.00

MADE IN GERMANY
ARMAND MARSEILLE
390
A. 6 M.

Socket head. Walker, head
turns, throws kisses.
22" $325.00

MADE IN GERMANY
ARMAND MARSEILLE
390
D.R.G.M. 246/1
A. 6½ M.

"My Dearie" for Geo. Borgfeldt
in 1908 to 1922. Socket head.
21½" $210.00 29½" $400.00

MADE IN GERMANY
ARMAND MARSEILLE
390
A. 6½ M.

Socket head walker. Straight
legs, head turns.
24" $325.00

390
ARMAND MARSEILLE
GERMANY
D.R.G.M. 240/1
A. 6½ M.

"Educational Doll" for Louis
Amberg. 1916. (Alphabet skirt).
Socket head.
21½" $210.00

390
D.R.G.M. 246/1
A. 6½ M.

"My Dearie" for Geo. Borgfeldt
in 1908 to 1922. Socket head.
24" $225.00

MADE IN GERMANY
ARMAND MARSEILLE
390
D.R.G.M. 266/1
A. 6½ M.

"Dora" for Butler Bros. in
1904. Socket head.
21½" $210.00

MADE IN GERMANY
390
A. 7 M.

Socket head.
23" $215.00

D.R.G.M. 377439
MADE IN GERMANY
D.R.G.M. 374830
A. 7 M.
390

Socket head. Fur eyebrows.
22" $235.00

390
A.M. 246/1
A. 3 M.

"My Dearie" for Geo. Borgfeldt
in 1908 to 1922. Socket head.
22" $210.00

ARMAND MARSEILLE
GERMANY
390
A. 7 M.

Socket head.
22" $210.00

MADE IN GERMANY
ARMAND MARSEILLE
390n D.R.G.M. 246/1
A. 7 M.

"My Dearie" for Geo. Borgfeldt
in 1908 to 1922. Socket head.
22½" $265.00

MADE IN GERMANY 390 A. 7 M.	Socket head. 22½" $215.00 30" $400.00
MADE IN GERMANY 390 A. 7/0 M.	Socket head. 9½" $95.00
MADE IN GERMANY ARMAND MARSEILLE D.R.G.M. 246/1 390n A. 7½ M.	"My Dearie!" for Geo. Borgfeldt in 1908 to 1922. Socket head. 24" $275.00
ARMAND MARSEILLE 390 A. 8 M.	Socket head mechanical ballerina on stand. (Electric). 1900s. 35" $750.00
MADE IN GERMANY 390n D.R.G.M. 246/1	"My Dearie" for Geo. Borgfeldt in 1908 to 1922. Socket head. 23" $265.00 27" $350.00
A. 8 M. 390 D.R.G.M. 24 A. 8 M.	Socket head. 35" $600.00
ARMAND MARSEILLE GERMANY 390 A. 8 M.	All composition socket head. 24" $185.00
ARMAND MARSEILLE 390 A. 8 M. D.R.G.M. 216 MADE IN GERMANY	In 1910, "Possy," a baby. In 1928, "Clara," a child. 10" $110.00 23" $215.00
ARMAND MARSEILLE 390 A. 8/0 M. D.R.G.M. 216	"Possy" socket head baby. 1928. 23" $215.00
ARMAND MARSEILLE 390 A. 8 M. RD.R.G.M. 216 MADE IN GERMANY	"Clara" socket head. 1928. 23" $215.00
ARMAND MARSEILLE 390 1 A. 9 M. D.R.G.M. 246 GERMANY	"My Dearie" for Geo. Borgfeldt in 1908 to 1922. Socket head. 24" $225.00 35" $600.00
D.R.G.M. 17 377439 MADE IN GERMANY D.R.G.M. 374830/374852 A. 9 M. 390	"Wonderful Alice" for Geo. Borgfeldt. Socket head. fur eyebrows. 24" $250.00 30" $450.00
ARMAND MARSEILLE 390 A. 9 M. D.R.G.M. 246 GERMANY	"My Dearie" for Geo. Borgfeldt in 1908 to 1922. Socket head. 25" $235.00 36" $650.00
MADE IN GERMANY ARMAND MARSEILLE 390 A. 9 M. ARMAND MARSEILLE 390	Socket head. 24" $225.00 40" $900.00 Socket head. Painted bisque. 9" $65.00

A. 9/0 M.
GERMANY

ARMAND MARSEILLE 390 A. 10 M. GERMANY	Socket head. 25″ $235.00
ARMAND MARSEILLE MADE IN GERMANY 390 A. 10/0 M.	Socket head. 8″ $65.00
MADE IN GERMANY ARMAND MARSEILLE 390n D.R.G.N. A. 10 M.	Socket head. 27″ $350.00
ARMAND MARSEILLE GERMANY 390 A. 11/0 M.	"Jutta" (on body) for Cuno and Otto Dressel in 1908. Open/ closed mouth. 7½″ $65.00 19″ $185.00
MADE IN GERMANY ARMAND MARSEILLE 390 D.R.G.M. 2½ A. 11/0 M.	Socket head. 7½″ $65.00 17½″ $175.00
MADE IN GERMANY ARMAND MARSEILLE 390 A. 11/0 M.	Socket head. 8½″ $85.00
ARMAND MARSEILLE GERMANY 390 A. 11 M.	Socket head. 26″ $250.00
390 D.R.G.M. 246/1 A. 11 M.	"My Dearie" for Geo. Borgfeldt in 1908 to 1922. Socket head. 26″ $250.00
D.R.G.M. 377439 MADE IN GERMANY	"Wonderful Alice" for Borgfeldt. Socket head. Fur eyebrows. 26″ $275.00
D.R.G.M. 374830/37481 390 A. 11 M.	Socket head. 26″ $250.00
MADE IN GERMANY ARMAND MARSEILLE 390n A. 12 M.	"Louisa" 1915. Socket head. 37″ $700.00
GERMANY A. 12 M. 390n	"Louisa." 1915. Socket head. 27″ $350.00
ARMAND MARSEILLE GERMANY 390 A. 12/0 M.	Socket head. 7″ $50.00
ARMAND MARSEILLE GERMANY 390 A. 12/0x M.	Socket head. 7½″ $55.00
ARMAND MARSEILLE 390 A. 13 M.	Socket head. Walker, head turns. 29″ $400.00

ARMAND MARSEILLE 390 A. 13 M. D.R.G.M. 246/1	"My Dearie" for Geo. Borgfeldt. 1908. Socket head. Voice box with pull strings. 30" $450.00
GERMANY 390 13/0	Socket head. 6½" $50.00
A.M. 390n	Socket head walker, cries as legs move. 24" $325.00
MADE IN GERMANY ARMAND MARSEILLE 390n A. 13 M.	Socket head. 28" $375.00
MADE IN GERMANY ARMAND MARSEILLE o. 3051/12 390n D.R.G.M. A. 13 M. 246/1	Socket head. Pierced ears into head. "My Dearie" for George Borgfeldt. 30" $500.00
ARMAND MARSEILLE GERMANY 390 A. 14 M.	Socket head. 29" $350.00
ARMAND MARSEILLE 390 A. 15 M.	Socket head. 30" $400.00 40" $900.00
395 GERMANY A 11/0 M.	"Heidi" for Geo. Borgfeldt Socket head. 1920. 9" $135.00
GERMANY 398 A.M.	Closed mouth Oriental baby. 10" $485.00
ARMAND MARSEILLE GERMANY 400 A. 2 M.	"Gibson Girl" and "Lady." Socket head. Closed mouth. For Louis Wolfe and Co. 1913. 18" $1,300.00
ARMAND MARSEILLE GERMANY 400 A. 3 M.	"Gibson Girl" and "Lady." Socket head. Closed mouth. For Louis Wolfe & Co. 1913. 18" $1,300.00
ARMAND MARSEILLE GERMANY 400 A. 5 M.	"Gibson Girl" and "Lady." Closed mouth. Adult. 1913. 13" $795.00
ARMAND MARSEILLE GERMANY 401 A. 5/0 M.	"Gibson Girl" played by Beverly Bayne (With Francis X. Bushman). 1913 for Louis Wolfe. 13" $795.00 17" $1,200.00
GERMANY 402 A 5/0 M.	Painted bisque socket head. 14" $235.00
448 A. 13 M.	Socket head. Painted bisque and mache. 28" $325.00
449	No information.
A. 450 M. GERMANY 1½	Socket head. Closed mouth. Dressed in Provincial costumes. 19" $850.00

A.M. 500 D.R.G.M. GERMANY	"Infant Berry." 1908. 8" $300.00 14" $1,000.00
500 A.M.	"Infant Berry." 1908. Mache socket head. Closed mouth. 18" $1,350.00
500 GERMANY A. 2 M. D.R.G.M. GERMANY	"Infant Berry" 1908. 12" $650.00 18" $1,350.00
500 A. 2/0 M. D.R.G.M.	"Infant Berry." 1908. Molded hair, dimples. 10" $450.00
500 GERMANY A. 3/0 M. D.R.G.M.	"Infant Berry." Intaglio eyes. 1908. 8" $300.00 20" $1,500.00
MADE IN GERMANY ARMAND MARSEILLE 500a A. 4/0 M.	"Infant Berry." Molded hair. 1908. 10" $450.00 24" $1,900.00
MADE IN GERMANY ARMAND MARSEILLE 500 A. 4 M. D.R.G.M. 1	"Infant Berry." Molded hair. 1908. 14" $1,000.00 18" $1,350.00
MADE IN GERMANY 500 A. 11/0 M.	"Infant Berry." Molded hair. 1908. 5" $150.00 8" $300.00
GERMANY 517 5K	Colored socket head. 20" $425.00
A.M. GERMANY 518 1K	Socket head toddler. Molded hair. White: 12" $195.00 Black: 12" $295.00
GERMANY A.M. 518 2½ K	Socket head. Baby. 1911. White: 26" $425.00 Black: 26" $625.00
A.M. GERMANY 518 3K	Socket head. Molded hair baby. White: 14" $265.00 Black: 14" $365.00
A.M. GERMANY 518 7K	Socket head. Baby. 1911. White: 19" $325.00 Black: 19" $425.00
A.M. GERMANY 518 8K	Socket head. 1911 baby. White: 21" $375.00 Black: 21" $475.00
A.M. GERMANY 518 16½ K	Socket head. 1911. White: 26" $425.00 Black: 26" $625.00
520 D.R.G.M. A. 2 M.	Socket head. Painted eyes. Open mouth. 15" $280.00
GERMANY 523 A. 2/0 M.	Socket head. Closed mouth googly. 9" $750.00 12" $1,100.00

GERMANY 550 A. 2/0 M. D.R.G.M.	Socket head. Closed mouth. 12″
GERMANY 550 A. 3 M. D.R.G.M.	Socket head. Closed mouth. 18″ $1,300.00
GERMANY 550 A 4/0 M. D.R.G.M.	Socket head. Closed mouth. 8½″ $350.00 16″ $1,200.00
GERMANY 550 A. 6 M. D.R.G.M.	Socket head. Closed mouth. 16″ $1,200.00
ARMAND MARSEILLE 560a D.R.G.M. 232	"Dorothy" 1912. Also 1924 toddler. 15″ $350.00 25″ $600.00
MADE IN GERMANY ARMAND MARSEILLE 560a A. 2M. D.R.G.M. M.R. 232/1	"Dorothy" 1912. Toddler and baby. Socket head. 1924. 9″ $185.00 15″ $350.00
MADE IN GERMANY ARMAND MARSEILLE 560a A 3/0 M. D.R.G.M. 232/1	"Dorothy." Socket head. 1924. 11″ $250.00
MADE IN GERMANY ARMAND MARSEILLE 560 A. 3 M. D.R.G.M. 232/1	"Dorothy." 1924. Bent leg baby socket head. 15″ $350.00 19″ $485.00
MADE IN GERMANY ARMAND MARSEILLE 560 A. 5/0 M. D.R.G.M. R 232/1	Socket head baby. "Dorothy." 1924. 9½″ $185.00
MADE IN GERMANY ARMAND MARSEILLE 560 A. 6 M. D.R.G.M. 232/1	"Dorothy." 1924. Socket head. 18″ $425.00 25″ $600.00
MADE IN GERMANY ARMAND MARSEILLE 560a A. 8/0 M. D.R.G.M. 232/1	"Dorothy." 1912. 8″ $175.00 20″ $485.00
MADE IN GERMANY ARMAND MARSEILLE 560 A. 5 M. D.R.G.M.	9″ $185.00 26″ $600.00
A.M. 580	Socket head. 15″ $225.00
590 A. O M. GERMANY D.R.G.M.	Socket head. Baby. 10″ $350.00 20″ $1,300.00
590 A. 5 M. GERMANY D.R.G.M.	"Hoopla Girl" for Hitz, Jacobs & Co. 1916. Open/closed mouth. 16″ $1,000.00 20″ $1,300.00

600 A.M. GERMANY D.R.G.M.	"Child Berry." 1910. Flange and socket head. Painted eyes. closed mouth. 10" $350.00 20" $1,400.00
600 GERMANY A. 2 M.	"Child Berry." Closed mouth socket head, painted eyes and hair. 1910. 14" $850.00
600 A. 3/0 M GERMANY D.R.G.M.	Shoulder head. Closed mouth, painted eyes. 8" $250.00 17" $1,100.00
600 A. 4/0 M. GERMANY D.R.G.M.	Socket head. Closed mouth, painted eyes. 10" $350.00 12" $500.00
600 A. 30 M. D.R.G.M. GERMANY	Shoulder head. Closed mouth. Painted eyes. 10" $350.00 25" $1,700.00
620 3	1925. "My Dream Baby" for Arranbee. All bisque. 8" $165.00
620 3/0	"My Dream Baby" All bisque. Closed mouth. 8" $265.00
ARRANBEE 620/4 GERMANY	"My Dream Baby." Flange. 7" $150.00 20" $425.00
640 A. 3 M. GERMANY D.R.G.M.	"Bernadette." 1909. Shoulder head. Painted eyes. 22" $385.00
ARMAND MARSEILLE GERMANY 690 A. 5½ M.	Socket head. 26" $550.00
ARMAND MARSEILLE 00 A. 9 M. GERMANY	"Beatrice" 1912. Closed mouth. Socket head. 14" $450.00 24" $1,000.00
ARMAND MARSEILLE GERMANY 700	Socket head. Closed mouth. 14" $450.00 24" $1,000.00
A. 9 M. 701	Socket head. Closed mouth. 14" $465.00 21" $950.00
ARMAND MARSEILLE 750 D.R.G.M. 257 A. 2 M.	Socket head baby. 1916. 14" $265.00 20" $425.00
A.M. 753 GERMANY	All bisque baby. 4" $125.00 10" $285.00
ARMAND MARSEILLE GERMANY 760 A.M. 5/0 D.R.G.M. 258	Shoulder head. 16" $325.00
A.M. 800	"Baby Sunshine" for Louis Wolfe. 1925. Also a "Mama" talker in head. 16" $300.00

917
A.M.

"Mobi" 1921.
Socket head baby. For Herman
Schiemer.
13" $250.00

A.M.
917
GERMANY

"Mobi"
Socket head baby/1921.
For Herman Schiemer
16" $300.00

ARMAND MARSEILLE
GERMANY
920
A. 5 M.

Shoulder head.
18" $365.00

ARMAND MARSEILLE
GERMANY
920
A. 6 M.

Shoulder head.
19" $365.00

A.M.
925
16" $325.00 20"

"Baby Bobby" for Geo. Borgfeldt
1926. Socket head.

A.M.
957
DEP.

Shoulder head.
18" $300.00

A.M.
966 3
MADE IN GERMANY

Socket head. Flirty eye
baby.
14" $285.00

A.M.
966 8
MADE IN GERMANY

Socket head, flirty eye baby.
20" $425.00

A.M.
966

Bisculoid socket head. Four
piece body. After 1910.
20" $225.00

GERMANY
970
A. O M.
D.R.G.M.

"Lady Marie" for Otto
Gans. 1916.
11" $150.00

OTTO GANS
GERMANY
970
A. 5 M.

"Lady Marie" for Otto Gans.
1916.
20" $250.00

ARMAND MARSEILLE
A.M. 971

"Minnit Baby" for Geo. Borgfeldt
Socket head. 1910
10" $150.00 16" $295.00

971
A. O M.
D.R.G.M. GERMANY

"Minnit Baby" 1910.
Socket head.
14" $165.00 25" $550.00

971
A. 1 M.
D.R.G.M. GERMANY

"Minnit Baby."
Socket head. Baby. 1910.
10" $150.00 16" $295.00

971a
A. 1 M.
D.R.G.M. 267
GERMANY

"Minnit Baby."
Socket head. 1910.
12" $250.00 24" $485.00

971
A. 2 M.
D.R.G.M. 267

"Minnit Baby."
Socket head. 1910.
White: 10" $150.00 16" $295.00
Black: 10" $200.00 16" $395.00

971
A. 2 M.
D.R.G.M. 267

"Minnit Baby."
Socket head. 1910.
10" $150.00 20" $425.00

GERMANY
971
D.R.G.M.
A. 2/0 M.

"Minnit Baby."
Socket head baby. Flirty
eyes. 1910.
11" $200.00 14" $285.00

GERMANY
971
A. 2/0 M.
D.R.G.M. 267/1

"Minnit Baby"
Socket head. 1910.
10" $150.00 18" $375.00

ARMAND MARSEILLE
971
A. 3 M.

"Minnit Baby."
Socket head. 1910.
12" $250.00 26" $550.00

GERMANY 971
A. 4/0 M.
D.R.G.M. 267/1

"Minnit Baby."
Socket head baby and toddler. 1910.
11" $200.00 14" $285.00

GERMANY
971
A. 5 M.
D.R.G.M. 267/1

"Minnit" for Geo. Borgfeldt in
1910. Socket head baby. Voice
box. 10" $150.00 16" $295.00

GERMANY
971
A. 6 M.
D.R.G.M. 267-1

"Minnit Baby."
Socket head. 1910.
White: 12" $250.00 20" $425.00
Black: 10" $200.00 20" $525.00

ARMAND MARSEILLE
A. 975 M.
GERMANY

"Sadie" for Louis Wolfe. 1914.
Socket head baby.
16" $295.00

ARMAND MARSEILLE
A. 975 M.
GERMANY
2/0

"Sadie." Socket head baby.
Flirty eyes. 1914.
9" $150.00

ARMAND MARSEILLE
A. 975 M.
GERMANY 3

"Sadie." Socket head baby.
Flirty eyes. 1914.
13½" $245.00

OTTO GANS
GERMANY
975
A. 5 M.

"Sadie." Socket head. 1914.
20" $325.00

OTTO GANS
GERMANY
975
A. 5/0 M.

"Sadie." Socket head. 1914.
11" $200.00

OTTO GANS
GERMANY
975
A. 7 M.

"Sadie." Socket head. 1914.
$17" $295.00

OTTO GANS
GERMANY
975
A. 7/0 M.

"Sadie." Socket head. 1914.
9" $150.00

ARMAND MARSEILLE
A. 975 M.
GERMANY
10

"Sadie." Socket head. 1914.
25" $500.00

ARMAND MARSEILLE
A. 975 M.
GERMANY
12

"Sadie." Socket head. Flirty
eyes. 1914.
23" $425.00

ARMAND MARSEILLE
A. 975 M.
GERMANY
13

"Sadie." Socket head. Flirty
eyes. 1914.
24" $450.00

GERMANY
977
A. 5 M.

Socket head.
15″ $195.00

A.M.
980
GERMANY

Socket head. Baby.
14″ $250.00

GERMANY
980
A. OM.
D.R.G.M.

Socket head. Toddler.
11½″ $200.00

A. 980 M.
GERMANY
9
D.R.G.M.

Socket head baby. Open/
closed mouth.
20″ $950.00

A. 980 M. 16
GERMANY

Mechanical. Eyes and tongue
move. 26″ $850.00

A.M.
985
GERMANY

Socket head. Baby.
12″ $200.00 26″ $550.00

GERMANY
A. 985 M.

Socket head. Baby.
15″ $295.00

GERMANY
A. 985 M.
2

Socket head. Baby.
15″ $295.00

GERMANY
A. 985 M.
3

Socket head. Baby.
13½″ $250.00

GERMANY
A. 985 M.
5

Socket head. Baby.
16″ $295.00

GERMANY
A. 985 M.
7

Socket head. Baby.
17″ $325.00 19″ $400.00

A. 985 M.
GERMANY 8

Socket head. Baby
12″ $200.00 26″ $550.00

ARMAND MARSEILLE
A.M.
990

"Happy Tot" for Geo. Borgfeldt
in 1910. Socket head. Flirty
eyes. Voice box.
15″ $295.00 19″ $325.00
27″ $650.00

ARMAND MARSEILLE
GERMANY
990
A 0½ M

"Happy Tot" 1910. Socket head.
13″ $225.00

ARMAND MARSEILLE
GERMANY
990
A. 2 M.

"Happy Tot"
Socket head baby. 1910.
14″ $250.00

ARMAND MARSEILLE
GERMANY
990
A. 2½ M.

"Happy Tot"
Socket head baby. 1910.
15½″ $295.00

990
A 3/0 M.

"Happy Tot"
Socket head baby. 1910.
10″ $150.00 20″ $325.00

990
A. 3/0 M.

Socket head baby.
8″ $100.00

ARMAND MARSEILLE GERMANY 990 A 3½ M.	"Happy Tot" Socket head baby. 1910. 16½" $295.00
ARMAND MARSEILLE GERMANY 990 A 4½ M.	"Happy Tot" Socket head baby. 1910. 16" $295.00
ARMAND MARSEILLE GERMANY 990 A. 7 M.	"Happy Tot." Socket head baby. 1910. 19" $325.00
ARMAND MARSEILLE GERMANY 990 A. 8/0 M.	"Happy Tot" Socket head baby. 1910. 7½" $100.00 11" $150.00
ARMAND MARSEILLE GERMANY 990 A. 8 M.	"Happy Tot" Socket head baby. 1910. 19" $325.00
ARMAND MARSEILLE GERMANY 990 A. 9 M.	"Happy Tot" Socket head baby. 1910. 20" $325.00
ARMAND MARSEILLE A. 11 M. 990	"Happy Tot." Socket head baby 1910. 21" $365.00
ARMAND MARSEILLE GERMANY 990 A. 12 M.	"Happy Tot" Socket head baby. 1910. 22" $375.00
990 A. 13 M. GERMANY	"Happy Tot." Socket head baby. 1910. 23" $400.00
ARMAND MARSEILLE A. 16 M. 990	"Happy Tot." Socket head baby. 1910. 27" $650.00
ARMAND MARSEILLE GERMANY 991	"Herbie" for Geo. Borgfeldt in 1922. Socket head toddler. 12" $200.00 18" $325.00
991 KIDDIJOY A.M. GERMANY	Socket head. 14" $265.00
991 GERMANY KIDDIJOY A. 7 M.	Shoulder head. Some have flirty eyes. 27" $550.00
A.M. 992	Socket head. 15" $250.00
A.M. 992-6	Socket head laughing baby. 1914. 23" $395.00
GERMANY 992 A. 5 M.	Socket head. Baby. 1914. 21" $365.00
ARMAND MARSEILLE GERMANY 992 A. 7 M.	Socket head baby. 1914. 22" $375.00

GERMANY
KIDDIJOY
JIO c 1926
993/3

Flange. Open/closed mouth. Molded hair.
19″ $650.00 22″ $950.00

ARMAND MARSEILLE
GERMANY
995
A. 2/0 M.

Socket head baby.
12″ $200.00

ARMAND MARSEILLE
995
A. 1½ M.

Socket head. Baby.
14″ $250.00

ARMAND MARSEILLE
GERMANY
995
A. 4½ M.

Socket head. Toddler and baby.
12″ $200.00 18″ $350.00

ARMAND MARSEILLE
GERMANY
995
A. 4 M.

Socket head toddler and baby.
18″ $350.00

ARMAND MARSEILLE
GERMANY
SUR
995

Socket head toddler or baby.
18″ $350.00

ARMAND MARSEILLE
GERMANY
995
A. 6 M.

Socket head toddler or baby.
27″ $650.00

ARMAND MARSEILLE
A. 10 M.
GERMANY
995

Socket head toddler or baby.
28″ $650.00

ARMAND MARSEILLE
GERMANY
996
A. 1 M.

Socket head.
16″ $295.00

ARMAND MARSEILLE
GERMANY
996
A. 2/0 M.

Socket head baby.
15″ $275.00

ARMAND MARSEILLE
GERMANY
996
A. 5 M.

Socket head baby or toddler.
17″ $325.00

ARMAND MARSEILLE
GERMANY
996
A. 6 M.

Socket head toddler or baby.
Clown: 17″ $425.00
17″ $325.00

ARMAND MARSEILLE
GERMANY
996
A. 7 M.

Socket head toddler or baby.
19″ $350.00

ARMAND MARSEILLE
GERMANY
996
A. 8 M.

Socket head toddler or baby.
Open/closed mouth, 2 molded teeth
18″ $550.00

997
GERMANY
KIDDIJOY
A. 3 M.

Socket head.
14″ $265.00

1330 A. 3 M.	Socket head. 14″ $165.00 18″ $225.00
KOPPLESDORF GERMANY 1330 A 2/0 M.	Socket head baby after 1910 for Heubach Kopplesdorf. 12″ $150.00 20″ $250.00
1347 FLORODORA	Shoulder head. Fur eyebrows. 21″ $295.00
3½ A. 1776 M. C.O.D. N. DEP GERMANY	Turned shoulder head for Dressel. 22″ $325.00
1804 A.M. 12/0 DEP. GERMANY	Socket head. 7″ $65.00
1980 A. 1 M. 372	"Banner Kid." ½ kid and ½ muslin. Shoulder head. 1895. 20″ $265.00
1894 A.M. DEP. GERMANY	Socket head. Black: 12″ $185.00 17″ $285.00 White: 12″ $150.00 17″ $210.00
1894 A.D.M. DEP. GERMANY	Socket head. Black: 10″ $165.00 20″ $325.00 White: 10″ $125.00 20″ $250.00
1894 A.M. O DEP. MADE IN GERMANY	Socket head. Black: 15″ $245.00 White: 15″ $185.00
1894 A.M. 1 DEP. MADE IN GERMANY	Socket head. Black: 16″ $265.00 White: 16″ $195.00
1894 A.M. 1½ DEP. MADE IN GERMANY	Socket head. Black: 16½″ $265.00 White: 16½″ $195.00
1894 A.M. 2/0 DEP.	Socket head. Black: 8″ $125.00 White: 8″ $95.00
1894 A.M. 2 DEP	Socket head. 17″ $210.00
1894 2/0 DEP.	"Cleonie" Socket head. Black: 12″ $185.00
1894 A.M. DEP. MADE IN GERMANY 2½	"Hindu Boy" Socket head. 17½″ $300.00
1894 A.M. DEP. MADE IN GERMANY 3/0	Socket head. 14″ $175.00
1894 A.M. 3½ DEP. MADE IN GERMANY	Socket head. 18½″ $225.00
1894 A.M. 4 DEP. MADE IN GERMANY	Socket head. Lever operated "Mama" talker. 19″ $385.00

1894 A.M. 4/0 M GERMANY	Shoulder head. 15″ $185.00
1894 A.M. 4/0x DEP. MADE IN GERMANY	Socket head. 11″ $135.00
1894 A.M. 6 DEP. MADE IN GERMANY	Socket head. 22″ $265.00
1894 A.M. 7/0 DEP. MADE IN GERMANY	Socket head. 12″ $150.00
1894 A.M. 8/0 DEP. MADE IN GERMANY	"Albert" Shoulder head. 1901. 13″ $165.00
1894 A.M. 8 DEP. MADE IN GERMANY	"Little Aristocrat" for Butler Bros. (paper tag). 1895. 23″ $285.00
1894 A.M. 9/0 M.	Shoulder head. 12″ $150.00
1894 A.M. 9 DEP. MADE IN GERMANY	Socket head. Toddler. 24″ $285.00
1894 A.M. 9 DEP.	Shoulder head. 24″ $285.00
1894 A.M. 10 DEP. GERMANY	Socket head. 24″ $285.00
1894 A. 11 M.	Shoulder plate. 26″ $300.00
1894 A. 11 M. MADE IN GERMANY	Shoulder plate 26″ $300.00
1894 A.M. 12/0 M GERMANY	Socket head. 13″ $165.00
A.M. 1895	Shoulder plate 20″ $250.00
1897 A.M. 4 DEP. MADE IN GERMANY	Shoulder head. 20″ $250.00
1897 A.M. 5 DEP.	"Bright Eyes" for Cissna in 1898. Shoulder plate. 12″ $150.00 18″ $225.00
A.M. GERMANY P.SCH. 1899 5/0	Shoulder head for Paul Schmitt. 16″ $195.00
A.M. 1899 6/0	Shoulder head. 11″ $135.00
1900 A. 10/0 M. 390	Socket head. 10″ $125.00
1900 A.M. MADE IN GERMANY	Socket head 18″ $225.00

1901 A.M. GERMANY	Socket head. 12″ $150.00 24″ $285.00
1908 W. DEP. C. 121 A.M. 1 A.M.	Shoulder and socket head. Pierced ears. For Welsch & Co. 16″ $225.00 24″ $300.00
1910 GERMANY SUNSHINE	Shoulder head. 24″ $285.00
A.M. 1910 SUNSHINE	Shoulder head. 25″ $300.00
2000 A.M. 4/0 DEP.	Socket head. 12″ $150.00
2015 QUEEN LOUISE	Socket head. 12″ $150.00 30″ $400.00
A.M. 2015 6 MADE IN GERMANY	Shoulder head. 22″ $265.00
A.M. 2549	Socket head. Painted bisque. 16″ $150.00
A.M. 2966	Socket head. Painted bisque. 18″ $200.00
GERMANY A.M. 2966 3 3/4 8	Socket head. 20″ $250.00
L.A.&S. 1914 G. 45520 2 GERMANY	"New Born Baby" for Louis Amberg. Flange. 12″ $250.00 15″ $300.00
L.A.&S. 1914 G. 45520 4 GERMANY	"New Born Baby" for Louis Amberg. 10″ $185.00 18″ $400.00
L. AMBERG & S. G 45520 GERMANY 4 1914	"New Born Baby" for Louis Amberg. 12″ $250.00 20″ $465.00
L.A.&S. 1921 G 45520 4 GERMANY	"New Born Baby" for Louis Amberg. 18″ $400.00 26″ $550.00
L.A.&S. 1921 G 45520 4 GERMANY	"New Born Baby" for Louis Amberg. 15″ $300.00
A.M. 2015	Socket head. 22″ $265.00
3200 A. O M. DEP. MADE IN GERMANY	Shoulder plate/head. Open/ closed mouth. 1895. Also turned head. 18″ $650.00
3200 O.A.M. D1	Shoulder head. Some turned. 14″ $195.00 21″ $285.00
3200 A.M. DEP.	Shoulder head. Some turned. 1898. 16″ $215.00 19″ $250.00
3200 A.M. 0½ DEP. MADE IN GERMANY	Shoulder head. Some turned. 1898. 14″ $195.00 20″ $275.00

3200 A.M. 1 DEP.	Turned shoulder head. 1898. 19" $250.00
3200 A.M. 3 DEP.	Shoulder head. Some turned. 22" $300.00
3200 A.M. 3/0 DEP.	Turned shoulder head. Felt body. 15½" $215.00
3200 A.M. 4 DEP.	Shoulder head. Some turned. 23" $325.00
3200 A.M. 5 DEP.	Shoulder head. Some turned. 15" $200.00
3200 A.M. 5/0 DEP. MADE IN GERMANY	Shoulder head. Some turned. 12" $150.00
3200 A.M. 6 DEP. MADE IN GERMANY	Shoulder head. Some turned. 25" $385.00
3200 A.M. 6/0 DEP. MADE IN GERMANY	Shoulder head. Some turned. 15" $200.00
3200 7 A.M.	Shoulder head. Some turned. 26" $400.00
3200 A.M. 8 DEP	Turned shoulder head. 22" $300.00 27" $485.00
3200 A.M. 8/0 DEP.	Shoulder head. Some turned. 13" $165.00
3200 A.M. 8/x DEP MADE IN GERMANY	Shoulder head. Marotte music doll (no arms or legs). 10" $350.00
3200 A.M. 9 DEP. MADE IN GERMANY	Shoulder head. Some turned. 24" $350.00
3200 A.M. 9/0 DEP. MADE IN GERMANY	Shoulder head. 11" $165.00
3200 A.M. 10/0x DEP.	Musical Jester. 1900. 10" $350.00
3200 A.M. 11 DEP.	Shoulder head. Some turned. 26" $450.00
3200 A.M. 11/0 DEP. MADE IN GERMANY	Shoulder head. 9" $140.00
3200 A.M. 12/0 DEP. MADE IN GERMANY	Shoulder head. 10" $155.00
A.M. 3300 7	Shoulder head. 23" $325.00
No. 3500 A.M. 1 DEP.	Shoulder head. 1900. 12" $175.00 26" $400.00
3500 A.M. 2/0 DEP. MADE IN GERMANY	Shoulder head. 17" $225.00

3500 A.M. 2 DEP. MADE IN GERMANY	Shoulder head. Some turned. 22″ $300.00
3500 A.M. 4 DEP. MADE IN GERMANY	Shoulder head. Some turned. 24″ $365.00
3500 A.M. 16/0 DEP. MADE IN GERMANY 3600	Shoulder head. 6″ $95.00
A.M. GERMANY 3524 7	Baby Gloria. Flange neck baby. 18″ $550.00
A.M. DEP. MADE IN GERMANY	Socket head. 16″ $225.00 26″ $400.00
A.M. O DEP. MADE IN GERMANY No. 3600	Socket head. 12″ $150.00
3600 A.M. 6 DEP MADE IN GERMANY	Socket head. 17″ $225.00
3700 A.M. 2/0 DEP MADE IN GERMANY	Shoulder head. Some turned. 18″ $225.00
3700 A.M. 3/0 DEP. MADE IN GERMANY	Shoulder head. 1901. 18″ $225.00
3740 30 FLORODORA 1374	Shoulder head. 21″ $285.00
1374 FLORODORA A. 12 M D.R.G.M. 3748-30	Shoulder head. 25″ $385.00 30″ $500.00
A.M. 4008 24″ $265.00	Shoulder head.
83115 A.M.	Socket head. 8″ $125.00 12″ $150.00
D.R.G.M. 201013 A. 2/0 M. GERMANY	Shoulder turned head. Talk mechanism in head. 16″ $295.00 20″ $400.00
A. 2½ M. D.R.G.M. 201013 MADE IN GERMANY	Turned head. Voice box. 24″ $500.00
D.R.G.M. 37749 A. 16 M. GERMANY	Socket head. Fur brows. 36″ $750.00
D.R.G.M. 115 774 39 370 D.R.G.M. 3748301 37483 A.M. 2 DEP. MADE IN GERMANY	Shoulder head. Some turned. 1898. Fur brows. 22″ $325.00